Dieter Helm is fellow in economi   or of economic
policy, University of Oxford. His previous books include *The Carbon
Crunch: How We're Getting Climate Change Wrong – and How to Fix It*
(2012, 2015) and *Natural Capital: Valuing the Planet* (2015), both published
by Yale University Press. In 2017 Professor Helm carried out the Cost of
Energy Review for the UK government.

'The book's grand scope and its provocative line should provoke considerable and fruitful debate.'
Peter Christoff, *Australian Book Review*

BURN OUT

THE ENDGAME FOR FOSSIL FUELS

DIETER HELM

UPDATED EDITION

YALE UNIVERSITY PRESS
NEW HAVEN AND LONDON

For information about this and other Yale University Press publications, please contact:
U.S. Office: sales.press@yale.edu   yalebooks.com
Europe Office: sales@yaleup.co.uk   yalebooks.co.uk

Typeset in Minion Pro by IDSUK (DataConnection) Ltd
Printed in Great Britain by Hobbs the Printers, Totton, Hampshire

Library of Congress Cataloging-in-Publication Data

Names: Helm, Dieter, author.
Title: Burn out : the endgame for fossil fuels / Dieter Helm.
Description: New Haven : Yale University Press, [2017] | Includes
   bibliographical references and index.
Identifiers: LCCN 2016043323 | ISBN 9780300225624 (c1 : alk. paper)
Subjects: LCSH: Energy industries. | Energy development—Environmental
   aspects. | Energy consumption—Environmental aspects. | Renewable energy
   sources. | Fossil fuels—Environmental aspects.
Classification: LCC HD9502.A2 H4549 2017 | DDC 333.8/2—dc23
LC record available at https://lccn.loc.gov/2016043323

A catalogue record for this book is available from the British Library.

IBSN 978-0-300-23448-0 (pbk)

10 9 8 7 6 5 4 3 2 1

To Sue, Oliver and Laura – as always

Contents

PART THREE
Creative Destruction and the Changing Corporate Landscape

Preface to the updated edition and acknowledgements

When you read this, the oil price could be anywhere between $20 and $100 a barrel. It could even be outside these boundaries. Although it will matter a lot to the companies, traders and customers, it will not tell you very much about the price in the medium-to-longer term. The fact that the price was $147 in 2008 and $27 in early 2016 just tells you that it is volatile. Bankers, investors and governments might get their fingers badly burned, but most will lick their wounds and survive another day if the falls since late 2014 are only temporary. But only if they are temporary.

From the perspective of our energy future, it is the trend and not the noise that matters. Until late 2014, there was a broad consensus about where prices were heading – up, ever up. Otherwise sensible and sane people convinced themselves that the world was running out of oil, and that demand was virtually insatiable from ever-growing China and the other developing countries in Southeast Asia, India and Africa. Constrained and then falling supply would collide with ever-rising demand, and there would be an economic shock making those caused by OPEC in the 1970s look tame.

Lest you think this is exaggerated, it is all there in the actions of the companies, the trail of quotes, statements and reports, and embedded in energy policies around the world, and especially in Europe. The oil companies were busily developing new resources at costs of up to or even above $100 a barrel. These ranged from the Arctic through to the tar sands in Canada. They put their money where their analysis had taken them.

Experts and institutions produced reports and books about 'peak oil' and the urgent need to diversify to protect customers and economies from the price shocks to come. Websites about peak oil abounded. To argue against peak oil in the mid-to-late 2000s was very much a minority sport and open to ridicule. It ran on well into the current decade, right up to the oil price collapse in late 2014.

The environmentalists and politicians bought into this narrative. They talked a lot about the coming shocks, and this played wonderfully into the hands of those lobbying for subsidies for nuclear and renewables. They could tell a story about how nuclear and renewables, though expensive now, would be in the money by around 2020, by which time oil – and especially gas – prices would have moved above the (high) costs of these low-carbon options. Nuclear in Britain was presented as a long-term bargain. Germany convinced itself that the *Energiewende* would be a good industrial strategy, capable of giving Germany a competitive (renewables) edge against the fossil-fuel-dependent US.

Companies dispensed large amounts on high-cost marginal investments; investors bought their shares; banks lent them lots of money; and energy customers were committed to paying for large-scale renewables programmes in offshore and onshore wind and first-generation solar. All any lobbyist needed was to come armed with a forecast of ever-higher fossil fuel prices.

It has all turned out very differently, at least in the short term. It may swing back again. Indeed it may already have done so by the time you read this. Or it may not. But this book is not about these short-term swings. It is about why fossil fuel prices may, in the medium and longer term, be heading gradually down; why, notwithstanding the price today or indeed on any given day, it is reasonable to expect the prices further out to fall. This is a book about the gradual demise of the fossil fuel industries, and how the transition will play out.

The end of fossil fuels is a comforting idea for many environmentalists, and in the end it probably will 'solve' climate change. But it is unlikely to play out in a neat way, or as a consequence of campaigns and political actions. We may end up leaving the superabundant fossil fuels in the ground, but it is unlikely to come about through boycotts, demonstrations and campaigns about 'stranded assets'.

The element of realism injected back into the markets from 2014 has been painful for companies, investors, renewables developers struggling to cope with cheaper gas, and for electric vehicles trying to compete with their oil-fuelled counterparts. But this is just a beginning. For the great oil-producing countries, such as Russia and Saudi Arabia, it is no picnic. It threatens the very survival of their autocratic regimes and the livelihoods of their citizens. So far, despite several attempts, they have failed to fix the price by credibly cutting supplies.

Russia is already struggling to pay for its overseas adventures in Crimea, eastern Ukraine and Syria. Saudi Arabia is having to start selling off parts of its crown jewel, Saudi Aramco, and confront the obstacles to modernization posed by its religious establishment. By contrast, the new energy world is a much better place for the US and Europe. Energy costs have fallen sharply, masked in Europe by all the levies for the legacy costs of first-generation renewables.

These impacts are the result of the ways in which the fossil fuels will be undermined. In the short run, it is all about the slowing demands of China's great economic expansion, which caused so much of the commodity super-cycle, as well as increasing the supply with the coming of shale and other new fossil fuel technologies. Further out, the challenge comes from new technologies and transformations in the structures of economies. This is a story about digitalization, the coming of robots, 3D printing, artificial intelligence, and the applications of communications technologies to infrastructures. It is therefore all about electricity – the electrification of almost everything – and how the generation, transmission, distribution and supply of electricity is changed by emerging generation technologies, electric cars, storage, batteries, distributed generation, smart grids, smart meters, and household and business broadband hubs. These technological advances are coming thick and fast.

This transformation of economies towards electricity – and the transformation of the electricity industry itself – changes almost everything for the oil and electricity companies. It changes their costs; it changes the nature of their markets; and it changes the competitive arena. Few if any of the big incumbents can look forward with confidence to much more than survival in the medium term – and certainly not to a ripe old age.

It is only with the hindsight of the historian, perhaps in 2050, that such trends and these medium-to-longer-term structural breaks with the past will become clear. We cannot know exactly how all this will pan out. There will be surprises. This book is about the ones that are, to an extent, predictable, and how the energy future might get radically changed. But it may turn out differently: indeed, there are bound to be new technologies that come along which will further change the game.

Technological change in my lifetime has been extraordinary. When I started out I was typing my thesis on an Olympia portable typewriter, which I carried around everywhere, much as I do now with my laptop on which I am writing this for you to read. There were no fax machines, no word processors, no Internet, no emails and no Google, Apple, Amazon or even Microsoft. What will the world of today's graduates look like in, say, thirty years' time? Try to imagine the main changes you think might be coming and see what you come up with.

In the energy sector, the resistance to the idea that the future might be very different from the past is endemic – and it almost always has been. Conventional oil and gas wells are much the same as they were fifty years ago. So are coal-fired power stations and even nuclear power stations. Transmission and distribution cables have changed little. Ask people working for the big energy companies and in energy policy to write down how they see things in thirty years, and it will probably not be very different, still with lots of oil, gas and coal, and a continuing gradual shift to current renewables.

Back in 2013 I was looking into the acquisition of a power station. In the way of these things, the investors paid for a couple of price forecasts to work out how good a deal it might be. They were exactly as I expected – an extrapolation of the past into the future. They were also in line with what the British government was forecasting in trying to work out whether the Hinkley Point nuclear power station would be a good deal, and how much customers should be forced to pay for the electricity it would generate over the next thirty-five years. The projections from the International Energy Agency looked very similar.

But to me this consensus looked wrong, and it was wrong. None of them had contemplated the possibility that the future might be a very different place. Energy modelling is not really about single-point forecasting, though investors and companies still like to have a base case to work from. When it is good, modelling is about exploring how significant changes feed through in markets and companies, and how the feedbacks and causal chains work out. They are more about 'what if' than 'what will be'.

To do this well, what is needed is a framework, and this requires much more than good data. The key is to get the questions right, and then the answers can follow. That means asking questions about broader economic factors, about the ways technologies change the nature of costs and hence the design of markets. When it comes to energy it is also about politics, governments and even war. This book is based on predictable surprises, grounded in developments that are already radically changing the energy world. At the core of this exercise are the following questions: What if prices fall rather than rise over the medium-to-longer term? What would the consequences be? Who wins and who loses? The answers often turn out to be surprising, even if they are also predictable.

Each could be the subject of numerous academic articles and books. Indeed, they probably will be. But I have not taken that approach here. These questions are vital to understanding what is in store for us in the twenty-first century and what our energy future will look like. In tackling them, I have opted for a broad approach – to elucidate how the answers fit together into an overall picture that is accessible not just to energy special-ists, but to the wider audience interested in global political and economic matters. No doubt the need to generalize will have offended specialists in each of the areas I tackle. But this is to miss the point of the book: I want my readers to think longer-term and generally. As a result I have also eschewed the practice of including lengthy and detailed references and endnotes, confining these to a few explanations and links to material which readers may want to explore further.

When it comes to writing a book such as this, the broader the canvas, the greater the influences and the debts owed to people who have been thinking about these issues. Trying to peer out into the medium and longer term, and across the energy industries, requires just such a broad canvas, and as a result there are many people who have helped, influenced and argued with me over recent decades.

First and foremost are my academic colleagues, without 'skin in the game' and hence able to take an independent and public-interest perspec-tive. Two stand out. Cameron Hepburn, the best mind I know on climate change and climate policies, has been a critical friend for many years, and his influence pervades this book, though he is not of course responsible for its many errors and should not be assumed to agree with my take on the substantive issues. He and I once planned to write a paper on 'predictable surprises', and this book is in part the contribution I would have made to that joint enterprise. Colin Mayer has been involved in one way or another with most of my work since we both started out in Oxford in the early 1980s. His influences on me are many and various. At Oxford, Chris Llewellyn Smith has been very helpful on the science side, as has Myles Allen. Alex Teytelboym's comments and criticisms have been invaluable. Malcolm McCulloch has helped with the chapter on technologies, including providing some very helpful data on solar photovoltaics. Matthew Bell has commented on the carbon chapter, and Edward Lucas has looked at the Russia chapter. Thanks are due to all of them.

Both Cameron and Colin worked with me in founding and developing Aurora Energy Research, alongside John Hood, Rick van der Ploeg and John Feddersen. John and Ben Irons have changed my thinking on many aspects of energy markets, and thanks are also due to Manuel Köhler, Florian Habermacher and Mateusz Wronski. Andreas Loeschel's knowledge of the German energy market has been invaluable. I have learned a lot from all of them.

Although it is fashionable in academic and some other circles to place little emphasis on the contribution of the managers of the great energy companies, this is a serious mistake. Decisions matter and there are choices as to the paths to follow. I am privileged to have known many of these managers and watched at close hand how they have handled both the economics and the politics over the past three decades. Let me pick out a few for special thanks: John Browne at BP, Ed Wallis at PowerGen, David Varney at British Gas, Sam Laidlaw at Centrica, Iain Conn at BP and now Centrica, Vincent de Rivaz at EDF, Keith Anderson at Scottish Power, Johannes Teyssen at E.ON, Peter Voser and now Ben van Beurden at Shell, and Helge Lund and now Eldar Saetre at Statoil, Charles Berry, now at Weir Group, Andrew Duff, now at Severn Trent, and Steven Holliday and now John Pettigrew at National Grid. Thanks in particular are also due to Chris Anastasi, Edward Beckley, Richard Abel, Gordon Parsons, Neil Angell, Richard Clay, Tom Crotty, James Flannagan, Janine Freeman, Angela Hepworth, Matthew Knight, Andrew Mennear, John Moriarty, Ben Noble, Cordi O'Hara, Peter O'Shea, Nick Park, Tom Restrick, Martin Stanley, Mark Shorrock, Mark Somerset, Lars Sørensen, Rupert Steele, Sara Vaughan and Jens Wolf.

I have worked with many ministers and officials in the energy sector over the years. I was Special Adviser to Günther Oettinger at the European

Commission in 2011, working on the 2030 energy and climate packages, and got to know many excellent Commission officials, of whom two stand out: Jos Delbecke and Peter Vis. Peter also greatly helped with his detailed comments on the Europe chapter. In Britain, I have known, one way or another, every energy minister since 1979, and most of their opposition shadows. These have included: David Howell, Nigel Lawson, Peter Walker, Cecil Parkinson, Tim Eggar, Michael Heseltine, Tony Blair, John Prescott, Margaret Beckett, Patricia Hewitt, John Hutton, Chris Huhne, Edward Davey, Amber Rudd and Greg Clark. For much of the period since the early 1980s there has been one minister roughly per annum, and a similar turnover on the opposition side. Among the recent special advisers, I have particularly benefited from discussions with Guy Newey, Stephen Heidari-Robinson and Josh Buckland.

In 2017 I undertook the Cost of Energy Review for the UK government, and developed a number of the themes of this book, notably in the design of electricity markets. Jeremy Allen at the Department for Business, Energy and Industrial Strategy led a great supporting team.

Daniel Russo has worked through the entire draft of this book, made pertinent and piercing observations and criticisms, and helped with the graphs and charts. Clever and practical, he has had a big impact on the book. Help on the charts for the updated edition was provided by Rowan von Spreckelsen. Earlier research assistance was provided by Nevena Vlaykova.

Putting the book together has been greatly assisted by Jenny Vaughan and Kerry Hughes, and Taiba Batool at Yale University Press has steered the project through to completion.

I am grateful to all of them, and of course, as every author knows, the family has to put up with the preoccupation that stretching one's mind across the canvas of a book necessitates. Thanks as ever to Sue, Oliver and Laura for their patience.

Figures

Abbreviations

AC	alternating current
AGR	advanced gas-cooled reactor
AI	artificial intelligence
bcm	billion cubic metres
Btu	British thermal unit
BWR	boiling water reactor
CAFE	Corporate Average Fuel Economy
CCGT	combined-cycle gas turbine
CCS	carbon capture and storage
CEGB	Central Electricity Generating Board
COP	Conference of the Parties
DC	direct current
E&P	exploration and production
EIA	US Energy Information Administration
EPR	European pressurized reactor
EU ETS	European Union Emissions Trading System
FBR	fast breeder reactor
FGD	flue-gas desulphurization
FiT	feed-in tariff
FSB	Federal Security Service
GDP	gross domestic product
Gt	gigatonne
GW	gigawatt

IEA	International Energy Agency
IMF	International Monetary Fund
INDC	Intended Nationally Determined Contributions
IOC	international oil company
IT	information technology
ktoe	kilotonne of oil equivalent
kWh	kilowatt-hour
LNG	liquefied natural gas
M&A	mergers and acquisitions
mbd	million barrels per day
mtoe	million tonnes of oil equivalent
MW	megawatt
MWh	megawatt-hour
NGL	natural gas liquids
NGO	non-governmental organization
NIOC	National Iranian Oil Company
NOC	national oil company
nTPA	negotiated third-party access
OECD	Organisation for Economic Co-operation and Development
OPEC	Organization of Petroleum Exporting Countries
PLO	Palestine Liberation Organization
ppm	parts per million
PWR	pressurized water reactor
QE	quantitative easing
R&D	research and development
RBMK	reaktor bolshoy moshchnosty kanalny
rTPA	regulated third-party access
SAGD	steam-assisted gravity drainage
SMP	system marginal price
SMR	small modular reactor
SOCAL	Standard Oil of California
SPD	Social Democratic Party of Germany
SRMC	short-run marginal cost
SSE	Scottish & Southern Electricity
SUV	sports utility vehicle
TANAP	Trans-Anatolian Pipeline

TAP	Trans-Adriatic Pipeline
TPC	Turkish Petroleum Company
UAE	United Arab Emirates
UNFCCC	United Nations Framework Convention on Climate Change
WTI	West Texas Intermediate

Introduction

Fast-forward to 2050 – almost thirty-five years from now. What will the world look like? How will technology have transformed our daily lives? Will it be a world of robots and artificial intelligence (AI)? Of graphene, fusion and electric transport? Now rewind – back to 1980. This was still a world of typewriters and the fixed phone line. No Internet, no apps, no mobiles, no laptops; not even any word processors (as we know them) or fax machines.

These long time horizons matter in energy. The future world will certainly need a lot of energy. That we can be confident about. We can also reckon that many of the decisions made today about energy will shape this world. Many of the power stations on today's energy networks were there in 1980, and many of those that were not were in the planning stages. The youngest coal power station in Britain started to come on-stream in 1974. The transport systems today look very similar, as do the nuclear power stations.

The bulk of our energy systems remain in place thirty-five years on: coal, gas and nuclear power stations; the internal combustion engine; oil exploration and large oil and gas wells; OPEC (Organization of Petroleum Exporting Countries) and the dominance of the Middle East; and Russia's vast reserves. There have been changes, the most important of which in terms of scale have been in the fossil fuels. The biggest has been the coming of gas: it was illegal in the US and Europe to burn it in a power station until 1990. Now gas-fired power stations compete directly with coal for market share. Then there is shale gas and oil, and the great advances in extracting fossil fuels. Wind farms and solar panels, biocrops to produce ethanol, and biomass are all 'new', but none has yet made much impact. For all these developments, overall the world depends on coal even more now than it did in 1980, and oil has not been toppled from its dominance of transport.

The world is still divided up into its main fossil fuel suppliers: the US, Saudi Arabia and Russia (all producing over 10 million barrels of oil per day, mbd), plus the rest of OPEC – and its main customers: China, Europe, Japan and the US again – and the long tail of the rest. With the exception of the US, the producers are overwhelmingly authoritarian, and, remarkably, many still assume that they will continue to become more powerful (and richer) as the supplies peak, with consumers having to continue to beg, condone and even invade as their economies are increasingly threatened by their rising energy costs. It is remarkable just how little has changed on the energy front in the last thirty-five years. Indeed, many of today's energy fault lines go back at least to the beginning of the twentieth century and the origins of the oil industries in Russia and the US.

The companies have reflected this stability. The gradual erosion of the market shares of the big private oil companies (international oil companies, IOCs) in favour of the growing band of ever more powerful state-owned rivals (national oil companies, NOCs) had begun in the 1970s and gradually played out in the 1980s. The oil industry is now largely in the hands of companies like Saudi Aramco, the National Iranian Oil Company, Kuwait Petroleum Corporation, Pemex and Rosneft, as well as Chinese companies such as PetroChina. Even in the democracies, companies like Statoil are largely state-owned. Over 90% of global reserves are in state hands, with the likes of Exxon, Chevron, Shell and BP forced to the periphery.

When looking backwards yields a picture of such continuity, it is not surprising that the future is seen as an extension of this fossil fuel past: oil for transport and petrochemicals, and coal for electricity and industry. Energy is what facilitated the great transformation of industry, and led to a world capable of supporting 7 billion people, compared with fewer than 2 billion in 1900. Fossil fuels have facilitated almost all the economic activity that has taken place in human history, freeing us of the constraints of very limited manual labour and horse power (literally) by opening up the huge capacity of the energy stored in the carbon-based fuels. The fossil fuels are what made the twentieth century possible.

The temptation to extrapolate this past into the future, and to see 2050 as a modified version of today, is almost overwhelming. Take the recent energy outlooks from the big companies. Despite lots of hype about the challenges ahead, and especially about decarbonization, Exxon, BP, Shell and Statoil all have oil, gas and coal playing a big part in the energy mix for the next half-century at least. The International Energy Agency (IEA) takes a similar view. Projections of coal-burn in power stations suggest that, whilst its recent climb from 25% to 29% of world primary energy demand between 1990 and 2013 might come under a bit of scrutiny, it will remain the dominant fuel for electricity for the next half-century. Oil is predicted to remain key to transport, with gas taking up some of the petrochemical demand and accounting for a good proportion of heating.

These projections flatly contradict the sorts of scenarios which would be required if significant climate change is to be averted and the 2015 Paris Agreement's ambition of limiting the temperature increase to just 1.5°C is to be achieved. Indeed, recent studies have shown that the global electricity generation capital stock has already built in a 2°C increase.[1] No more carbon-based investment can be made if the target is to be met. Thus there is a basic and fundamental disconnect between the assumptions of business-roughly-as-usual and the fate of the planet. The stakes could not be much higher.

For all the force of the negative impacts of temperatures rising above 1.5°C or 2°C that scientists warn us about, the world appears to be heading ostrich-like in the direction that governments and companies project. After a quarter of a century of trying, following the 1992 UN Framework Convention on Climate Change (UNFCCC), and despite repeated 'pledges', 'targets' and 'global agreements', emissions (and the stock of carbon in the atmosphere that they contribute to) just keep going up. Only economic crises and the slowdown of Chinese growth have made any significant impact, and these may be temporary. The favoured alternatives, such as the current solar panels, wind farms and biofuels, have not made any serious difference other than to raise electricity costs and reduce competitiveness. Nor can they ever make enough difference, sadly. There simply is not enough agricultural land for energy crops, sites for wind turbines on land and in shallow water, or rooftops for conventional solar panels for such intermittent and low-density sources of electricity to make any real inroads, except locally.

Are we therefore doomed to repeat the twentieth century in the twenty-first? Is it business-as-usual with wind farms and solar panels bolted on (and lots of subsidies for corn and rapeseed oil production)? Must we confront the inevitable chaos and bloodshed in the Middle East, recognizing that we will be forever held to ransom by rich autocratic oil states? Must the Europeans bow down to President Putin and his successors, grateful for Russian gas supplies? Must China develop a blue-water fleet to defend its oil supply route through the Strait of Hormuz, and must we all start building sea walls to adapt to inevitable climate change?

These are among the biggest questions of our age. The answers are of course uncertain, but it is not all doom and gloom. Fortunately, the conventional wisdom is based on much shakier foundations than either the oil producers or the big oil companies would have us believe. This time it really may be different. Why? Because there are three big 'predictable surprises' out there, which together will transform their – and our – world, economically and geopolitically.

The three predictable surprises are: the end of the commodity super-cycle, and with it the gradual fall in oil and gas prices into the medium and long term; the carbon crunch, as the climate change realities dawn; and the wide-ranging revolution that is going on in technology. Ever-greater supplies and gradually falling demand will cut away at the fossil fuel prices, reinforced by carbon constraints and competition from new energy sources, mostly electric.

Following the abrupt price falls in late 2014, there are good reasons for thinking that there will be no return to the commodity super-cycle any time soon. Oil prices at $40–$60 per barrel (or even lower) may be the new normal. Prices may revert to the pattern of the 100 years between 1870 and 1970 – a remarkably stable trend, with prices gradually declining. The two exceptions, 1972–80 and 2005–14, may turn out to be aberrations, and not the norm.

Prices may fall even lower: in the medium term because there is much more production to come from Iran and Iraq, and potentially further major increases in supply as shale technologies go global and existing reserves are much more intensively depleted; and in the longer term, because demand for oil may go into a gradual decline as the new technologies cut into transport and petrochemical demand.

More pressure will be exerted through increases in carbon prices (explicitly through carbon taxes and permits, or perhaps more likely, implicitly through regulations) to offset the falls in oil prices, driving a greater wedge between the price of fossil fuels and the final energy prices consumers face. It may not be enough to choke off demand, but it will make a difference, and especially to coal – indeed, it already is. Add in direct regulation of the multiple emissions from coal, and it may well find itself replaced in part by gas. The fossil fuels may remain dominant, but the mix will turn out to be very different. This has already come as a 'surprise' to those coal companies that have suffered a drop in their share price, or even bankruptcy (such as the US company Peabody), and to the electric utilities that rely on coal (such as Germany's RWE, which has seen its finances decimated).

Technological progress on a scale not seen in energy for over a century is the enormous elephant in the room that is already transforming energy, with radical implications for the big companies, OPEC and the rest of the producers, and for geopolitics. It is not just one specific technology, and it is not a single silver bullet. It is a revolution that touches each and every part of energy production and consumption. And it is the best hope of tackling climate change.

One massive technological advance – fracking – has already transformed the fossil fuel industry, changed geopolitics, brought new companies into the market and halved the oil price. It has taken only eight years to produce these dramatic impacts. The combination of horizontal drilling, new seismic-information technologies, and the ability to split open rock struc- tures has turned the US from a declining oil and gas producer, with ever- rising costs and imports, back into a renewed fossil fuel superpower, as it was until 1970. Astonishingly, in less than a decade, the US added 3 mbd, became the world's largest oil and gas producer, and set itself firmly on the path to (roughly speaking) energy independence by the 2020s. All of this is due to technological progress. Policy has played no significant part.

The shale revolution is a *revolution*: it has turned conventional assump- tions on their head, destroyed the myth of 'peak oil', dealt a massive blow to OPEC, and helped to dramatically reduce the price of oil on international markets. In turn, it has irrevocably altered the prospects of Putin and Russia, undermined the post-Chávez government in Venezuela, and thrown all the main producing countries in the Middle East into deficits, including Saudi Arabia. Saudi Arabia has had to borrow from the international debt markets and might even have to introduce serious taxation. Incredibly fast, turning conventional assumptions on their head, and dramatically changing the numbers – this fits any definition of a revolution.

Revolutions take time to play out. The fallout of this one will go on for decades. But if oil supply is abundant and demand gradually softens, countries like Saudi Arabia will have to recognize that the happy assumption that oil produced tomorrow will be worth more than that produced today is not robust. The new energy abundance threatens authoritarian regimes relying on their natural resources. Low prices helped to bring down the Soviet Union in the late 1980s/early 1990s, and Russia to its knees at the end of the 1990s. Neither Gorbachev nor Yeltsin could withstand the resulting loss of revenues. Subsequently, rising prices underpinned Putin's power after 2000 through to 2014 – but no longer.

The really deep, fundamental energy revolutions do not, however, lie with specific fossil fuels, like shale oil and gas. Shale has merely revealed what is obvious to all but the peak oil brigade who thought that ever-higher prices would make wind farms and current solar panels economic. The uncomfortable fact is that the earth's crust is riddled with fossil fuels, enough to fry the planet many times over. As prices rise, so too do the incentives for technical innovations. If the resources are there, economic incentives usually work in finding new ways to extract them. And they have – spectacularly.

The really revolutionary surprises lie in more general technological progress and they are largely about electricity. Electricity is increasingly the energy source of choice. Gradually, through digitalization, we are moving towards the Internet of Things, in which almost everything is electric. Any process that is digitalized is electric. Electricity will transform transport, the core of the current oil demand. Electrifying transport will be a revolution in its own right. Electricity will also transform heating and cooling.

So what matters for the future of energy is how the electricity is generated. Electricity generation is wide open to technological change. Opening up the light spectrum and developing new ways of capturing the energy of the sun through new materials and onto solar film together offer opportunities that no fine-tuning of wind turbines or existing solar panels could ever achieve. The future of electricity is probably solar, but not as we know it.

Electricity can be stored. Household batteries are already being installed. Car batteries might become to the electricity industry what petrol tanks in cars and trucks are to oil storage. New information technologies are already transforming the demand for energy, from smart apps controlling central-heating systems to smart grids and smart meters. Then there are new cable technologies, which might make very long-distance electricity transmission economically feasible, bringing solar energy from the Sahara and geothermal energy from Iceland into the European energy mix.

How these technologies pan out and mesh together in the energy systems of the future is obviously hard to predict. Indeed, it would be spurious to attempt to pick specific winners and simply predict the future on the basis of these sorts of assumptions. That is the mistake politicians keep making, especially in Europe. Claiming to know the future in a fine-grained way is beguiling, but it has not been, and probably never will be, a successful strategy. The radical technologies are almost always 'general-purpose technologies', not specific discoveries. The Internet, and before it the coming of the railways, cars and electricity, were just such general-purpose technologies, changing the very nature of economies *generally*. What we have already is the knowledge of the possibilities and some of their general characteristics, but not much more.

Recognizing uncertainty does not mean we are completely ignorant, or that we need be paralysed. From an energy perspective, one crucial economic fact stands out. Almost all of these new electricity generation technologies do not have an energy cost once they are built and installed. The energy they generate is free. In the technical jargon, it has zero marginal cost. In this respect it is like the Internet and broadband. The equipment is expensive, but it costs nothing to generate the electricity, just as it costs nothing for me to use the Internet once the systems are in place. The systems have to be paid for, just as solar power has to be paid for. But once installed, the cost of an additional unit of electricity generation is zero.

Getting to grips with this idea means turning almost everything we know about the electricity industry on its head, and since the future is electric, this applies to the energy sector generally. Current energy markets are based on the idea that the main driver is the positive variable energy costs. Variable marginal costs are the main driver of wholesale markets, and wholesale markets are where prices are determined. The oil price is based on a marginal variable cost, as is the wholesale electricity price. There is a unit price for electricity and indeed all the fossil fuels. It might be $50 per barrel for oil, and $30 per megawatt-hour (MWh) for electricity.

A zero-carbon world is close to a 'zero marginal cost electricity generation' one. Since there is no substantive wholesale market, almost all the economic action happens through fixed-price contracts – because the costs are overwhelmingly fixed and sunk too. It is a world of capacity contracts, feed-in tariffs (FiTs), and of fixed monthly customer bills. Like the monthly deals for broadband, it is paid for by an access charge for the use of the system, not a use charge, even if it is dressed up as if the volume of demand matters. It is not a world of liberalized wholesale markets, which the architects of the great experiment of the last two decades had in mind. It is *radically* different.

A zero marginal cost world is one where the growth of demand does not much matter. As long as the energy is low-carbon, who cares about reducing energy demand (a current obsession among politicians)? Maximizing demand might be a better strategy as long as the value of the use of the energy is above zero, just as maximizing Internet use is overwhelmingly a 'good thing'. Nobody advocates minimizing broadband use. Imagine politicians trying to persuade people to ration their use of the Internet. In a zero marginal cost world, electricity could potentially be even 'too cheap to measure'. That, after all, is what zero marginal cost means. There is just capacity, and once installed, usage is not really important. No more policies of reducing energy demand. Who pays for the capacity, and especially the burden on the fuel-poor, is an open question.

The demand side is at the same time being transformed to be active rather than passive, to the extent that this matters at all. But there is yet more technical change on the way. It is not just energy *directly* that is changing; it is also the structure of the economy itself and the spatial distribution of manufacturing. 3D printing is radical in that it undermines the very idea of mass production. The ever-larger factories that dominated twentieth-century economies may be a thing of the past. 3D printing allows for bespoke production, just-in-time and highly localized. Crucially, it is not necessarily globalized.

Robots are able to do much of the stuff factory workers did during the last century, and not just the routine activities. Robotic production lines are now the norm in car factories. Robots do not sleep or demand wages and welfare. They can be active and intelligent. The cost of labour and the associated competitive advantages of China, India and Southeast Asia are no longer so relevant. The energy needs of a 3D printer production economy are quite different from those of Ford's Model T production lines, or of Apple's iPhone factories in China. For the energy future, this means not only that everything is electric, but that the assumption that manufacturing is inevitably shifting from Europe and the US to China and the developing world (with its cheap labour) gets turned on its head. The location of energy demand may therefore be very different. The notion of relentless globalization may be reversed. Why produce in China and import when you can produce at home for the same cost?

Perhaps just as dramatic are the new materials, of which graphene is one exciting example. It is a single layer of carbon atoms, with great electrical conductivity properties, and is both extremely strong and flexible. Some have suggested that it is as important as the discovery of plastics. Time will tell. Imagine if graphene and other new materials did actually replace plastics. Imagine what the consequences would be for petrochemicals – the other main use for oil, after transport. If petrochemicals were confronted with a serious rival material, Saudi Arabia's problem with the US's abundant shale and potential energy independence would be greatly exacerbated, further challenging the country's monarchy and religious foundations. Shale, opening up the light spectrum, solar film being widely applied, electric cars, new batteries solving the electricity storage problem, information technologies activating the demand side, new material like graphene, a 'Second Industrial Revolution' with robots and 3D printing – these are the things that OPEC, Middle Eastern monarchs and autocrats, Putin and oil executives should worry about. These developments could make the current falls in oil prices look like a picnic in comparison. They are what our energy future may be all about.

They are also all key parts of an effective strategy to deal with climate change. Existing technologies won't solve the problem. They can't. Instead of spending ever-greater sums of money, and ever-more political capital, on expensive technologies like wind farms and current-generation solar panels, and diverting land from the production of crucial food supplies to making ethanol and biofuels, it would be much better to channel funds towards these new technologies. That would make much more climate and economic sense.

Although governments have done little if anything to help these technological revolutions along, and have wasted billions on things that cannot make a big difference, it is impossible to hold back the tide of what is coming in the next few decades. The oil producers and the oil companies are in for repeated shocks, as were the old telecoms companies, the old supermarkets and the old publishers, as the Internet and the associated communications revolution got going. But the oil producers and oil-rich companies are not hopeless flotsam on the tide of technology: they can change *if* they recognize that the existing economic models are not sustainable.

Step one is to acknowledge the possibility of radical change and of discontinuity. The world's geopolitics is predicated on the existing patterns of energy demand and supply. The companies are based on the existing cost structures, which in turn are driven by the characteristics of the existing technologies. Tomorrow, they assume, will be roughly like today. The starting point is to recognize that if the underlying economics – the facts – change, the existing models will be increasingly exposed to competitive challenge. What should Saudi Arabia do, faced with abundant fossil fuels and the new technologies? What should Exxon, Shell and Statoil do, faced with the zero marginal cost, low-carbon technologies? As the facts change, they should change their minds and their models.

It is reasonably safe to predict what most of the incumbents will actually do: very little initially, carrying on as usual. That is what IBM did, and what the telecoms companies like BT did, and as a result the future was not theirs but rather that of Microsoft, Apple, Google, Amazon and Facebook. The chances are that the twenty-first century does not belong to the existing players: their days are probably numbered. But this is not inevitable: IBM did eventually reinvent itself, as did several of the telecoms companies like BT, though their dominance was comprehensively broken. RWE and E.ON have both failed to hold on to their dominance in Germany, yet E.ON has started to reinvent itself, and even RWE, after trying most of the other alternatives, has begun to recognize the new realities, though not before halving its asset value.

When it comes to governments and countries, many of the current energy giants may have to contend with returning to their 'former littleness', a fate William Stanley Jevons forecast for England in 1865 as he predicted that it would run out of coal in the second half of the nineteenth century.[2] Saudi Arabia could gradually morph back into a relatively insignificant desert state (though it would need to work out what to do with its much greater population). Even its sovereign wealth fund is unlikely to sustain the shock of ever-lower prices. Indeed, just a year on from the price collapse in late 2014, it had already suffered serious damage. Consultant-driven mega-plans to transform its economy away from oil dependency might work, but the odds are heavily stacked against the Saudis. Others with a broader economic base, like Iran, may fare better. However, as a resource-based economy for the last few centuries, Russia's fate looks bleak since it has few competitive industries beyond fossil fuels, other than its military complex.

On the other – and more positive – hand, the US holds many of the aces, especially technological depth. It has energy abundance, cheap gas and the world's greatest technological capabilities. The energy future is more likely to be American than Chinese. China might build its navy to defend its oil and gas supply routes, but it may not ultimately need them. The threat to China's model comes from the loss of competitive advantage that the new manufacturing technologies may bring. Its energy requirements will change as a consequence, not as a cause. The great Chinese abundant-and-cheap-labour model has already run out of steam. Ironically, just when cheap fossil fuels might help China's slowing economy, new technologies are threatening to undermine its economic advantages, and have begun lowering global trade and triggering a process of reshoring. It is not clear whether the country can successfully transition its economy to meet these daunting challenges.

This book directs the spotlight on the great energy break with the twentieth century and on what the combination of the new normal fossil fuel prices and the technological revolution might mean for the major energy countries, for the companies and for the climate, and considers what the future energy markets might look like. It will of course be wrong – there will be *unpredictable* surprises as well as predictable ones – but a great deal can be envisaged. Being wrong is inevitable. What matters is being interestingly wrong, since the future is what we determine.

PART ONE

Predictable Surprises

As US Secretary of Defense Donald Rumsfeld famously remarked in the context of the Iraq War and the alleged weapons of mass destruction, there are 'known knowns', 'known unknowns' and 'unknown unknowns'.[1] The first are facts, the second are predictable surprises, and the third are just surprises. This book is largely about the second.

Predictable surprises are those things that we can reasonably expect to happen. They may not of course: predictions are just predictions, not facts. Yet they are the material for thinking about the future of energy, and about how this future will impact on the producer countries, the companies and the consumers. They should be incorporated into the design of company strategies and energy policies.

There are many predictable surprises. The challenge is to pick out those that are most likely to matter, to shape the energy sector as a whole, and thereby impact on geopolitics and the corporate landscape.

Of the many possible candidates three stand out. The first is the end of the commodity super-cycle of the last decade, dominated by the rise of China, and its subsequent return to 'normal'. Re-normalized, oil prices may fall over the medium and longer term. The second is the carbon constraint: the climate change agenda and its scientific support will eventually force out the fossil fuels. Both of these predictable surprises are reinforced by the third: the coming of new technologies. Each will be sufficient to effect a significant break with the past. The three together will result in a radical transformation and the gradual endgame for fossil fuels.

The three chapters that follow take each in turn, and provide the context for considering the impact on the main fossil fuel producers and consumers in Part Two, and on the companies in Part Three.

CHAPTER 1

The end of the commodity super-cycle

The first predictable surprise has already happened: the end of the commodity super-cycle and with it the crash of oil, coal and now gas prices from the end of 2014. This collapse came as a surprise to many, but it was predictable. What comes next depends on understanding why the super-cycle happened in the first place, and why it collapsed. On this the future path of commodity prices turns.

Those who think the lower prices in 2015 and 2016 are an aberration, and that the cycle will be back with a vengeance, hold out the prospect of continuity: continuity of OPEC and Russian behaviour. For them the crash is a blip – painful, but ultimately temporary. Once the high-cost producers are forced out it will be business-as-usual all over again. Lower prices will do their work: supply will come down and demand will go up again.

For those who think that there are fundamental economic forces at work which might reinforce the price falls and extend them into the medium and longer term, the collapse is the initial episode of a long drama which will change the very nature of energy markets, undermine the power of the oil producers, and change the nature of the companies. Supply and demand will do their work in this story too. Prices may well go up again in the short term through to 2020, but not in the medium and long term. The latter view is the one this book focuses on.

Commodity cycles are not new

The belief in the permanency of specific commodity super-cycles is just another manifestation of the gullibility of investors, company executives and politicians. The seductive argument that there are 'special reasons' why prices can go only one way is not unique to energy. It has been believed by investors in the South Sea Bubble (1720), in Dutch tulips (1800), in the run-up to the Wall Street Crash (1929) and most recently in the sub-prime crisis (2007–08). There is no evidence that this susceptibility is anything other than a fundamental part of human nature.

What all these bubbles have in common is the associated belief that the laws of economics can be suspended. Yet the prices of commodities are not exempt. Whilst the prices will not always be the efficient ones (and they rarely are in oil), supply and demand do have to equate – a bit like double-entry bookkeeping. For every seller there has to be a buyer. Someone has to pay – and they must be both willing and able to do so. When the price goes up, demand falls and supply increases. If the price increase is sustained then technical innovation is encouraged. These are the economic facts of life.

This is precisely what has happened in energy, and why the price of oil halved in late 2014, and then fell even further in 2015. It was a very predictable surprise. These simple bits of economics explain much of what has been going on in energy markets over the last couple of decades – indeed over the last half-century – and yet they pass most politicians, and many company executives, by.

A cursory look back at oil price predictions in the run-up to the 2014 collapse is sobering. Goldman Sachs expected $200 per barrel.[1] The major companies all publish economic outlooks, and these too make embarrassing reading. The key point is that their chief executives *believed* them, and they backed up these forecasts with billions of dollars of investments which depended on high prices, from the tar sands in Alberta to drilling in the Arctic. Analysts encouraged this. Just four months before the crash, leading oil market academic, James Hamilton, summarized an analysis of the reasons why prices would stay up as follows: 'My conclusion is that hundred-dollar oil is here to stay.'[2] The politicians were perhaps just naive in talking up oil prices, about which they were generally pretty ignorant. The leaders of most of the main European countries were especially gullible, and they put their faith in these prices, assuming they would make the renewables they subsidized so enthusiastically economic by around 2020. They really did believe all this.[3]

In the energy world, for many the gold standard for analysis and future projections is the IEA. It makes the news headlines, and is quoted across the world whenever it opines on energy markets. Thus, when it comes to super-cycles, you might expect it to have a good track record. But it doesn't: its record is awful. It led the way on the 'higher-for-ever' bandwagon – just as it had in predicting ever-higher prices after the 1979 Iranian Revolution. It turns out that IEA forecasts (or 'projections' as it prefers to call them) have been so bad that that since the 1970s it has almost always been better to extrapolate the current price than rely on the IEA's expertise.[4]

My colleagues at Aurora Energy Research analysed the record. Figure 1.1 shows the actual oil price and the predictions (projections) made by the IEA at various points since the early 1980s. Aside from the sheer scale of the forecasting errors, the key insight from this chart is that the IEA *did not forecast the price falls* in the mid-1980s *or* in 2014–15 (even if it also missed

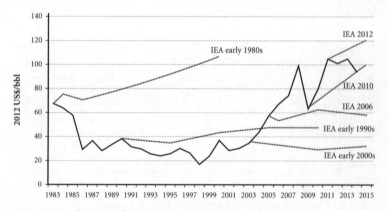

Figure 1.1 Crude oil price forecasts versus actual, 1983–2015, per barrel

Sources: Aurora Energy Research, 'Predictable Surprises: Lessons from 30 Years of Energy Sector Forecasts', 2013, with data from IEA, *World Energy Outlook, 1982, 1993, 2000, 2006, 2010, 2012*, © OECD/International Energy Agency, *World Energy Outlook*, IEA Publishing

the price rises in the late 2000s). The other widely respected global energy agency, the US Energy Information Administration (EIA), is not much better.

The errors are just too big to be random. Something much more fundamental is going on here. The IEA is generally asymmetric in its upward bias, and it is probably no accident that its role and budgets are better justified in a world of price shocks and associated security concerns. But that has not stopped companies and governments taking it extremely seriously.

Bringing these forecasts right up to date, the current downturn in prices was not anticipated, as Figure 1.2 shows. Indeed, at each point on the path of oil price declines, the IEA predicated a rebound. It was all going to be very temporary before things settled down and returned to 'normal'.

Whether it was because the oil producers and companies (and many analysts) believed the IEA, or because their own models relied on the same equations, they very much followed the IEA path in consistently failing to predict the falls. Figure 1.3 gives some examples of their predictions.

Futures markets were no better. These are where people put their money where their mouths are. Figure 1.4 shows that as the price fell from 2014, at

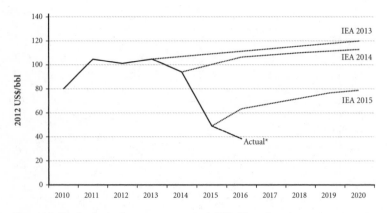

Figure 1.2 Crude oil price forecasts versus actual, 2010–20, per barrel

Note: * Average 2016 Brent crude oil price as of May 2016.

Sources: Aurora Energy Research, 'Predictable Surprises: Lessons from 30 Years of Energy Sector Forecasts', 2013, with data from IEA, *World Energy Outlook, 2013, 2014*, 2014 © OECD/International Energy Agency, *World Energy Outlook*, IEA Publishing

Figure 1.3 Oil price commentary

Sources: See endnote[5]

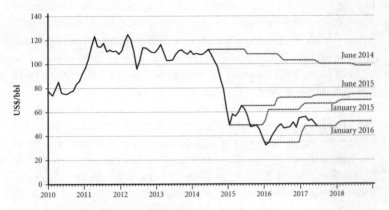

Figure 1.4 Crude oil futures prices versus actual

Source: Aurora Energy Research, with data from Thomson Reuters

every point on the price decline the futures markets projected a quick stabi-
lization. By January 2016 the futures markets suggested $40 in 2020. They
have since jumped around a lot.

Anything could happen in the next couple of years. The lesson from the
price crash after 2014 is that short-term markets can be very volatile, and

there can be big swings in prices on the basis of small changes in demand and supply. After such a price shock it is perfectly understandable that the market oscillates while trying to find a new equilibrium. Every bit of new data, from the latest Chinese manufacturing numbers, to the Federal Reserve's views of US interest rates, to meetings and deals between the Saudis and the Russians, is seized upon as an exhibit in the chaos of markets. But best to ignore these swings, and even stop reading the daily reports. Our energy future does not rest on any of these short-term behaviours. Whether the price goes up a lot in the next year, or falls further, tells us almost nothing about the medium and longer term.

The long view

History is a lot more useful. It takes the long view. Although people have always used oil and bitumen, the oil industry as we know it today is only 150 years old. The early development of kerosene, initially for lighting, began simultaneously in Baku (then in Russia, now the capital of Azerbaijan) and in Pennsylvania in the US. The new technologies for refining crude oil spread quickly once the internal combustion engine had been developed.[6]

With hindsight it is hard to appreciate now how little recognition there was in these early days of the huge potential for oil, or indeed the sheer scale of the resource base. Despite the prominent predictions of the exhaustion of supplies in virtually every decade since, throughout the twentieth century the supply of oil kept up with demand. There is nothing new about the 'higher-for-ever' predictions, despite the evidence to the contrary.

The widely used BP chart in Figure 1.5 shows the oil price from the 1860s through to today. It is a truly remarkable story. Despite two world wars, the sheer scale of economic development, the unprecedented growth in world population and the transformation of industry and transport, real prices remained pretty constant from 1880 until 1970. In fact, the overall trend is slightly downward: *prices actually fell over the century.* Supply responded to ever-increasing demand and incremental technical improvements reduced the costs. The history of the oil industry is to a considerable extent a history of these supply responses. The dotted line added to this chart shows the historical average over the entire period to give an indication of what might now be 'normal' over the long run.

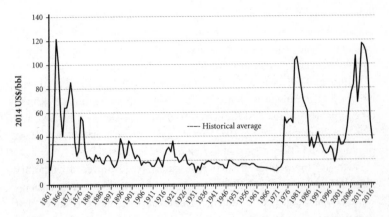

Figure 1.5 Crude oil prices, 1861–2016

Source: BP, 'Statistical Review of World Energy 2015', 64th edition, June 2015, bp.com/statistical review (historical average added)

After the initial finds in the late nineteenth century, other deposits were discovered and developed outside Russia and the US. Iran became an early additional key source, and led to the founding of the Anglo-Persian Oil Company (later becoming BP) in the early decades of the twentieth century. Venezuela was the other early major source of supply outside the US and Soviet Union. It was only after the Second World War that the elephant fields of Saudi Arabia were discovered and developed, transforming what had been a relatively small and insignificant tribal desert kingdom into a key player in world markets, with all the geopolitical consequences that went with its new status.

In the 1970s this stable pattern of supply responding to demand broke down, creating the first of the two great aberrations in the long-run trend (the second is after 2000, discussed below). With prices rising but marginal costs low, the allocation of the resulting economic rents became a major bone of contention between the companies and the producing countries. In the 1950s and 1960s the golden age of economic growth in Europe, the US and then Japan pushed up oil demand and they became politically and economically dependent on Middle Eastern oil. At nominal prices of $2–$4 a barrel, there had been little constraint on demand growth, and little incentive to look elsewhere for new sources of supply, given that it was (and indeed still is) so cheap to produce in the Middle East. Why develop more expensive supplies before exhausting the cheap ones first? While US production was close to a peak at the end of the 1960s, the wars with Korea and Vietnam added to its growing import dependency.

Although OPEC had already been set up in 1960, and the oil producers had tried to coordinate their negotiations to give more traction against the oil companies, it took first the Israeli-triggered Six-Day War in 1967 and then the Yom Kippur War in 1973 to really galvanize the key Middle Eastern producers, and OPEC then had its first decade in the sun.[7]

War and the politics of solidarity against Israel proved a temporary political cement for the Gulf States, and even Iran and Libya found themselves cooperating. The oil embargoes jolted the oil consumers out of their complacency, and by the end of 1974 oil prices had quadrupled to $11 a barrel. The scale of the shock, and the apparent inability to do anything about it, in turn created a broader crisis in capitalism. Inflation combined with unemployment to produce stagflation. Some, like Britain, were brought to their knees; a couple of years of more than 20% inflation and industrial unrest resulted in the need to turn to the International Monetary Fund (IMF) for assistance in 1976, followed by the 'winter of discontent' in 1978/79, when even the dead were not buried.

In the US, after President Richard Nixon fell in the Watergate scandal, the rest of the decade saw the hapless attempts by Gerald Ford and then Jimmy Carter to find a way out of the aftershocks of the price increases. For Carter, it was about finding alternatives like solar, energy efficiency, reducing demand and, most importantly, coal, as it was to be again in the late 2000s for the Europeans.

In the event, it just got worse. The Iranian Revolution of 1979 resulted in the oil price peaking at the nominal level of $39 (around $135 in 2015 prices). This was a further *doubling* of the price. It was enough to convince political leaders that the 1970s were not an aberration, but a fundamental structural break with the past, and they now had to confront a future of ever-rising prices.

There was, for Jimmy Carter,

> no doubt . . . that everywhere in the world, oil prices and general energy prices have been going up, and there is no doubt that in the future those prices will continue to go up.[8]

Germany's Chancellor Helmut Schmidt held a similar view:

> For the rest of the century oil prices will have to go up because oil reserves are gradually being used up.[9]

France's President Valéry Giscard d'Estaing agreed, as did the European Commission's President, Roy Jenkins.[10] All of them, after the experience of the horrible 1970s, could see only a world in which the oil price acted like the squeeze of a boa constrictor, leaving them powerless to cope with the immediate aftershocks.

Company executives agreed with the politicians' prognosis. Even though they cut their demand forecasts,[11] the prediction of ever-rising prices was widely assumed. Academic groups followed suit.[12] Since the political and business leaders were convinced of a future of increasing prices, policy was directed at finding alternatives to fossil fuels. The nuclear industry was a primary beneficiary, notably in France and Japan.[13] France's fifty-eight pressurized water nuclear reactors (PWRs) were born of these OPEC shocks, as were Japan's fifty-plus nuclear reactors. Both were explicit strategic responses to the 1973 embargoes and price shocks. In Britain, the Thatcher government announced in 1980 a programme of ten PWRs, with the aim of building one per year.[14] Germany, Belgium, Sweden, Italy and even Spain got in on the nuclear act too.

These projections (backed as we have seen by the IEA), and the certainty with which they were made, are eerily reminiscent of those made over the first decade of this century. The words used by these 1970s and early-1980s politicians could have been spoken by Obama, Merkel, Hollande and Barroso – encouraged again by the IEA and the EIA's confident 'projections'. Keep the quotes and policy statement, change the dates and the names, and there is an uncanny similarity in the understanding and responses in both cases. History may not repeat itself, but politicians and business leaders often do.

The peak oil delusion

Armed with the certainty of 'higher-for-ever', the new conventional wisdom in the 1970s and the 2000s found itself a theoretical rationale. It was called *peak oil*, and this in turn rested on a wider consensus in the 1970s that the world was incapable of sustaining its growing population and continued economic growth. The Club of Rome Report in 1972 purported to extrapolate demand against the 'known' supplies of a number of key commodities and food production, and predicted an inevitable Malthusian nightmare:

> If the present growth trends in world population, industrialization, pollution, food production, and resource depletion continue unchanged, the limits to growth on this planet will be reached sometime within the next one hundred years. The most probable result will be a rather sudden and uncontrollable decline in both population and industrial capacity.[15]

Peak oil was a seductive way of underpinning the certainty about rising prices. The building blocks on the supply side were the assumptions that all major reserves had been discovered and that depletion rates were known. Therefore, as demand kept on going up, it followed that supply would peak and then fall. Accordingly, M. King Hubbert, the father of peak oil, predicted that US supply would peak in 1970 and then decline.[16] Whilst his prediction turned out to be right (at least until shale came along), almost everything else about the peak oil hypothesis turned out to be wrong. But that did not stop those who already *knew* the future from using it as a convenient prop for their case. They have kept the peak oil faith alive, with a host of websites and commentaries 'explaining' each apparent failure.

There were vested interests at play here too. If the oil price was going ever upwards, and if a host of commodities were running out, alternatives would be needed and economic growth would be constrained. For some, this required a whole new approach to the organization of society in a zero-growth world. For others, it was the great argument for subsidizing their preferred technologies. The beneficiaries of these subsidies could advance the seductive argument that as the oil price went up, such subsidies would be only *temporary*, and hence the politicians were not committing to long-term support, but rather were prudently picking winners.

Yet, just when the certainty over prices was being enunciated with ever-greater conviction in the 1970s and early 1980s, the economics was undermining it. This 'certainty' raised the net present values of energy efficiency measures, and the demand side began (with a lag) to reflect the new conventional wisdom. Gas-guzzling Cadillacs gave way to new economy vehicles.

Consumers and industry used less of what was now a much more expensive commodity.

But if demand fell relative to trend, supply went up. Higher prices rendered a host of previously uneconomic resources worth pursuing, as they would again in the mid-2000s. Alaska and the North Sea were at the time the high-cost marginal resources. Alaska's Prudhoe Bay, the largest US oil field, was discovered in 1967, and West Sole (gas), Montrose and Forties in the North Sea were also all of 1960s vintage. They blossomed in the 1970s under the OPEC price stimulus, as shale oil and gas, deep-sea drilling and tar sands would in the mid-2000s. To these new resources were added greater supplies from OPEC itself. The new, much higher prices made cheating on OPEC quotas ever more attractive, and discipline naturally collapsed, increasing OPEC output. The proximate cause of the first aberration in the long-run trend – Arab unity – ceased to work, and has not re-emerged since.

By 1985 Saudi Arabia had had enough. The combination of new resources from areas like Alaska and the North Sea and cheating on quotas by the other OPEC members led to retaliation. Saudi's production had been absorbing the consequences. As Figure 1.6 shows, it fell back from 9 mbd in

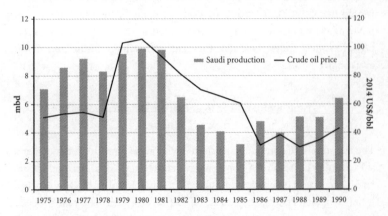

Figure 1.6 Saudi Arabia's crude oil production and price, 1975–90

Note: Figures include a share of production from the Neutral Zone.

Source: OPEC, 'Annual Statistical Bulletin (ASB) 2015', Organization of Petroleum Exporting Countries, http://asb.opec.org/index.php/data-download; Thomson Reuters

1979 to less than 4 mbd in 1985. Saudi Arabia decided (as it would again in 2014 and 2015) to defend its market share and open the taps. The oil price fell back, and remained low until the end of the century. It reverted to the trend of the previous 100 years.

The geopolitical results were spectacular. It was enough to materially help in finishing off the Soviet Union in the 1980s, and subsequently to undermine Yeltsin's post-Soviet attempts at reform in the 1990s (as we shall see in Chapter 6). The one exception was the brief price spike resulting from the First Gulf War in 1991 – and the remarkable feature about this episode is that, despite the damage wrought by Saddam Hussein to the Kuwaiti oil wells, the price spike was *very* short-lived.

Just as high prices led to reduced demand and increased supply, so low prices led to increased demand and some fall-off in the development of new supplies. They did not, however, kill off Alaska or the North Sea. The very large upfront costs of developing these new resources were sunk. Nor did the low prices reverse the energy efficiency gains.

After fifteen years, the market began to reflect the increases in demand and the slowdown in new resource developments. From a floor of around $10 a barrel in 1999, prices began what turned out to be a sustained rise. In 1999, $10 was a very low price compared with all those expectations back in 1979, especially taking into account the high inflation of the intervening two decades. It made a mockery of many of the investments that policymakers and executives had committed to at the end of the 1970s, based on their convictions that they *knew* the energy future and it was ever-higher prices.

Wrong again: failing to anticipate the phenomenal rise of China

The rise in oil price from 2000 reflected what was going on the demand side. The new big feature of global energy markets was China, and its phenomenal – and largely unanticipated – growth, partly spurred on by cheap energy. It was a major contributing cause of the second great aberration in the long-run trend. China's energy-intensive, export-orientated economic development gradually grew to dominate the world's energy markets. At 7–10% gross domestic product (GDP) growth per annum, the economy doubled in less than ten years, and then did it again and then again. Anyone doubting the scale of the impacts on global energy markets should reflect on what would have happened had the Chinese transition not materialized. Prices might have continued at around $10–$30 a barrel for the 2000s, and there would probably have been no shale or tar sands worth extracting.

It is important to get a handle on just how much extra energy demand this Chinese economic miracle represented. Figure 1.7 sets out China's oil and coal demand between 1980 and 2016.

The phenomenal economic growth of China represents one of the great economic transitions in world history, but like all transitions it works itself through and has now begun to come to an end, even if the official growth rate projections pretend otherwise. It is no accident that the end of the Chinese miracle is likely to coincide with the end of the great commodity super-cycle, and the end of the second great aberration in the long-run trend. Whilst a repeat of this phenomenal growth cannot be ruled out in India and parts of Africa, the chances of the emergence of the peculiar set of circumstances that conditioned China's transformation are slim, notably in a digitally driven rather than cost-driven global competitive process.

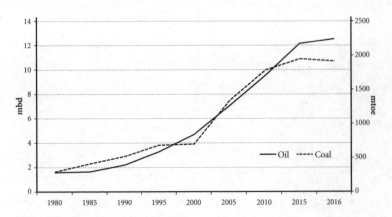

Figure 1.7 Chinese oil and coal consumption

Note: mtoe, million tonnes of oil equivalent.

Source: BP, 'Statistical Review of World Energy 2017', 66th edition, June 2017, bp.com/statistical review

Demand in developed countries: the continuing decoupling of
GDP from energy demand

China is the dominant story behind the recent commodity super-cycle, just
as it is behind its unwinding. But it is not the only story. The counterpart of
the rapid Chinese demand growth is what happened to US, European and
Japanese demand in the same period since 2000, as prices started to rise, as
shown Figure 1.8.

There are two parts to the great moderation of energy demand in these
countries – the point at which economic growth and energy demand apparently decoupled. The first is the changing structure of these economies, and
the second is the gains in energy efficiency.

The counterpart to the growth of China's energy-intensive exports is
the deindustrialization of these same industries in the developed countries.
As China built up its steel, petrochemicals, fertilizers and other energy-
intensive industries, the US and Europe reduced their shares in their
economies, and in some cases output actually declined. US steel, for
example, declined at nearly 1% per annum between 1975 and 2002, while
growing at over 10% per annum in China.[17] Industrialization in China
was matched by deindustrialization in the developed countries. US and

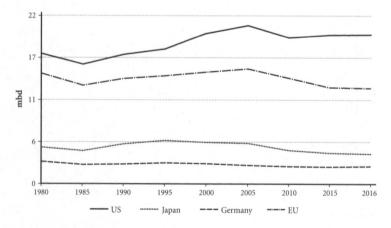

Figure 1.8 Oil consumption by region

Source: BP, 'Statistical Review of World Energy 2017', 66th edition, June 2017, bp.com/statisticalreview

European shares of manufacturing declined,[18] with services taking their place.

The second factor, energy efficiency, is largely a story about gradual incremental change, ratcheted up in response to higher prices. Vehicles get lighter, hybrid cars are encouraged, the internal combustion engine gets improved, insulation gets installed and machines get smarter. These gains are in addition to those that result from the direct demand response to higher prices. Importantly, these gains are irreversible.

The 'surprise' of shale and fracking

Whilst demand is gradually choked off by rising prices, and the recent exceptional trend in Chinese demand growth works its way back to normalcy, higher prices also induce supply-side responses – with an inevitable lag, as companies took time to realize that the uptick in prices after 2000 is likely to be sustained, and to implement their development and investment plans.

The impacts of the price rises since 2000 followed this traditional pattern: they encouraged development of marginal oil and gas fields, this time in deeper waters; they encouraged substitution to renewables; and they induced technical change, producing the spectacular explosion of shale and other unconventional technologies.

Just as the OPEC price shocks in the 1970s brought the first significant moves towards shallow offshore drilling, notably in the North Sea, so the gradual rises after 2000 led to the development of the frontiers of drilling, in particular to the deeper offshore. The big new developments concentrated on the Gulf of Mexico and the deep waters off Brazil and East Africa, and encouraged drilling to explore in the Arctic, the South China Sea and even off the Falkland Islands. Few of these would have been worthwhile at $10–$20 a barrel, but at over $100 a cornucopia of prospects were worth exploring.

Prospects also brightened in onshore areas which might have had too high a cost or political risk in the 1990s. As prices increased these marginal investments became more attractive. Lots of conventional sources in Africa were explored, and a number of companies (such as Genel Energy) carried on trying to get oil out in northern Iraq (with very mixed results), and even in a post-Gaddafi Libya (ENI). In Russia, despite Putin's very political elimination of Yukos, and the dubious auction which gave its assets to the state-controlled Rosneft, the oil sector has been transformed during the Putin years back to its former output levels.

What really changed the game was the response to higher prices in the US and Canada. It is a remarkable story. As Figure 1.9 shows, in less than a decade, fracking technologies in the US transformed the world's oil markets, added over 3 mbd to US output and, along with gas, catapulted what had been a declining oil and gas industry into the largest combined producer in the world – up there with Russia and Saudi Arabia. Canada's tar sands in Alberta added to the picture of the transformation of new fossil fuel technologies. Hubbert's 'peak' turned out to be temporary.

In one sense, the shale revolution is pretty straightforward and it could have happened at any time in the last quarter of a century. It is the combination of three technologies, each of which had been independently developed, in some cases decades earlier. These are: the ability to drill horizontally with precision; the IT and seismic information capabilities to understand the nature of rock structures and to analyse the data from the technology in the drill head; and the ability to split, or frack, the rock at the drill head.

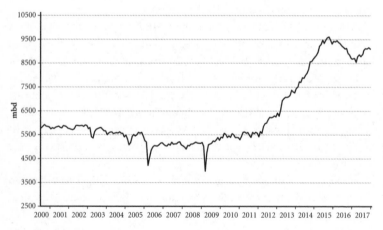

Figure 1.9 US crude oil production, 2000–17

Source: EIA, 'Crude Oil Production', August 2017, http://www.eia.gov/dnav/pet/pet_crd_crpdn_adc_mbblpd_m.htm

Fracking injects water, sand and chemicals to split open the rocks so the oil and gas can flow back to the wellhead.

The final ingredient to make fracking happen is access to land. Here the US has a number of key advantages over almost everyone else. The rights to the minerals below the surface rest in the US with the landowner, not the state. Landowners have a big incentive to make fracking happen, and the outcome is some very happy farmers. Incentives are aligned.

It helps too that the US has lots of big open spaces to do the drilling – a further contrast between the US and aspiring frackers in Europe. There might be great prospects in the Paris Basin, the southern counties of England and the industrial heartlands of Germany, but the combination of proximity to people and the pressure voters can put on local and national authorities to exercise their sub-surface property rights to prevent fracking are many and varied. Add to this the deep vested interests of the renewables lobbies with their subsidies to defend against the often much cheaper costs of reducing carbon emissions by using gas instead of coal, and in some countries the vested interests of the nuclear industries too, and a swift emulation of the US was never going to happen in Europe.

With many of the necessary background conditions in place, what made it happen in the US, and happen extremely fast, was the price. By the mid-2000s the price of oil was heading back from its low of $10 towards $100 per barrel, transforming the economics and kicking off a stampede to develop new resources fast. Add in very low interest rates and hence lots of cheap debt to finance the developments, this new combination of technologies encouraged massive 'learning-by-doing' effects, and the costs began to tumble.

Extraordinarily, it turned out that the costs of the fracked gas could be *lower* than for conventional production. Gas prices collapsed in the US, changing the global energy game, and changing the country's economic prospects. Figures 1.10 and 1.11 indicate how fast gas production rose, and how fast gas prices fell. Contrary to past experience, it all happened very quickly.

It wasn't just shale. With deposits of oil that rank in the top three in the world, Canada's tar sands offered another non-OPEC source of supply. The technologies could be basic – shovelling up the tar sands and sepa-rating out the heavy oils – or they could be sophisticated and build on the

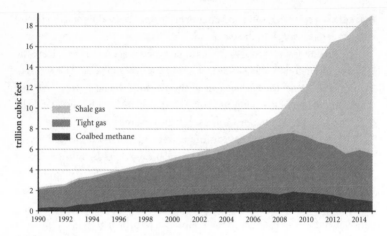

Figure 1.10 US unconventional natural gas production, 1990–2015

Source: EIA, 'Natural gas production by source in the Reference case, 1990–2040', *Natural Gas Annual 2017*, DOE/EIA-0131 (2017), Washington, DC, January 2017

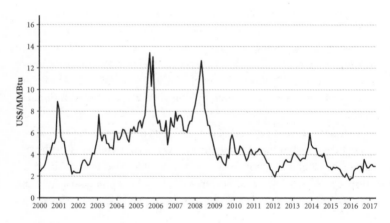

Figure 1.11 Henry Hub natural gas spot price, 2000–17

Source: EIA, 'Henry Hub Natural Gas Spot Price (Dollars per Million Btu)', September 2017, http://www.eia.gov/dnav/ng/ng_pri_fut_s1_m.htm

development of fracking. Steam-assisted gravity drainage (SAGD) may not sound exciting, but it opens up large deposits below the surface. Parallel horizontal drilling injects the steam at one level to soften up the bitumen to make the oil flow at another, up to a kilometre away from the drilling site. To get some idea of scale in the vast swamp-like conditions of northern Alberta, a single 'pod' might have ten such parallel drilling lines and there might be up to, say, ten pods per site. Each pod might cost \$5 billion. This is oil development on a vast scale.

The net results of these various developments have not only been to add enough extra supply to crack the world price, but also to undermine any remaining faith in the concept of peak oil. It is no longer plausible to assume that the stock of future oil reserves is known, that it is limited to the conventional wells, and that there is a limit of around 100–110 mbd in production that cannot be exceeded. Hubbert and his followers are just plain wrong. The earth's crust is riddled with shale rocks and carbon deposits. It is just a question of cost, price and whether it is worth investing in the technologies to get them out. Indeed the problem is not peak oil, or even peak demand, but rather (as we shall see in the next chapter) *peak carbon*: there is *too much* oil and gas left, as well as the vast (and from a climate perspective, especially dangerous) coal deposits.

Much more oil and gas to come

The peak oil theory runs into difficulty even where conventional oil is concerned. It assumes away technological progress and, in particular, progress in enhancing the level of depletion. Once a well is opened up, the oil comes out at pressure. Often this pressure is a real problem, as in the Deepwater Horizon disaster in 2010. It requires very strong control systems to avoid blowouts, explosions and uncontrolled spills. But as the well is depleted, the high-pressure problem is replaced by another: falling and eventually low pressure. There comes a point when it is no longer worth relying on a trickle, even after trying the usual ways of keeping the flow going – injecting water into the wells, for example. Getting just 50% out is a great achievement with the conventional technologies.

Think what this means. In existing and known wells there is more oil left than has been depleted in the entire history of the world's oil industry.

It is still there, but inaccessible. But what if the pressure can be kept up for longer? What if just 1% more is economically extractable? As the price rises, these last marginal reserves become increasingly attractive, and a host of technologies have been developed in the last fifteen years or so to get to them. There are at least three: drilling further wells at the margins; enhancing the pressure in the wells; and fracking at the peripheries of wells. All of these are currently being explored in the North Sea as the once-great oil wells reach their depletion thresholds. Importantly, like fracking more generally, they will not be un-invented.

Of these, enhanced recovery is particularly interesting. Instead of, for example, flooding wells with water, and in the process doing a lot of damage to the remaining reserves, an alternative option is gas injection, and in particular CO_2 from the burning of fossil fuels. Injecting gas back into the wells has many advantages over water. Better still for the oil companies, if governments can be persuaded that this is part of developing carbon capture and storage (CCS) technologies, then the economics may begin to stack up in certain locations. These costs are, however, very high.

The development of fossil fuel technologies, the growth of the non-OPEC supplies, and especially the growth of US shale production, have induced a sense of foreboding in the OPEC countries. The weakening price outlook is the trigger for the breakdown of cartel discipline and, as in the mid-1980s, Saudi Arabia has again led the charge. This time it did not need to increase production: it merely refused to cut output as the combination of weakening demand and increasing supply spilled out into the market. As we shall see in Part Two, the Saudi game is more than a simple short-term response to shale: it is about a longer-term market in a world where there is a chance that oil prices may stay low. It is as much a battle with Iran and Iraq as it is with the US and Canada about future market shares.

The scale of potential Iranian and Iraqi production contributes to this longer-term game. Although neither is likely to make a big difference to global oil supplies in the short term, their potential overhangs the market. Iran is more immediate. Iraq has plenty of potential too, and both Iran and Iraq are technically capable of producing over 10 mbd in the context of a total current world supply of around 90 mbd. We return to this in Part Two where we consider the future evolution of the Middle East.

It's all over: the end of the commodity super-cycle

As demand has fallen and supply gone up, the great commodity cycle has come to an end. The price falls have been remarkable, and often dramatic, across a range of commodities. The great Chinese boom for commodities is largely over; energy efficiency gains are permanently embedded into the markets; shale technologies are now a fact; and the renewables are growing.

What happened as the price fell was predictable, even if many of the players and commentators were surprised. The challenges facing major oil-producer countries are now at best uncomfortable and at worst close to existential. They relied on high prices, and they had spent the proceeds of the great super-cycle.[19] Venezuela was already in trouble. Now it was bust. Others had come to rely on $100 per barrel and they felt the pinch. Putin had spent almost all of the windfall, and the price fall plunged Russia's finances into the red, making the disastrous incursions into Crimea and Ukraine and the sanctions that followed much worse for Russia than they would have been during the commodity super-cycle. Even Saudi Arabia had already been spending up to the price ceiling, like many of the other authoritarian regimes, in part to buy off the revolutionary threats unleashed by the Arab Spring. We turn to the geopolitical implications of all this in Part Two.

What do oil-producing countries do when the money runs out? The answer in the short run is: *pump more*. More output at lower prices helps to prop up the revenues and hence the government budgets. Discipline within the OPEC cartel, to the extent that there is any, breaks down in a falling-price world. And it has: no one cut production in 2015 and into 2016. Last time around, in the mid-1980s, it took fifteen years for the price to bottom out, and then only because Chinese demand was starting to make inroads into supplies.

It is not just the countries that react to lower prices in an apparently perverse way. Companies are often financed on the basis of higher prices, and now they need the money to cover their short-term interest calls and dividends. Many of their costs are fixed and sunk in the development of the wells. So they keep pumping. US shale production proved remarkably resistant.

The dynamics of the market process take time to play out. Most appear to expect that, eventually, a higher cycle will kick in: lower prices will increase demand; and supplies will be choked off. In 2015 all the major companies cut their capital budgets, and cut their headcounts too. Investment in nuclear and renewables is less attractive, as it was in the 1990s, and indeed prior to the OPEC price spikes of the 1970s. They expected that up will go the price again. For the countries and companies it is just a matter of toughing it out. That indeed is what many commentators have suggested is Saudi Arabia's game. The prices at the end of 2015 are regarded as 'low', not 'normal', even if some had begun to suspect that they might be 'lower for longer'.

This time it may really be different. Or rather the cycle may be swamped by other factors outside the oil market's control. There are two further predictable surprises: the impact of carbon policies and carbon prices will limit the feed-through of lower oil prices into lower prices for industry and consumers, as carbon prices and emissions controls plug the gap (Chapter 2); and the nature and speed with which technologies outside the fossil fuel industries develop will undermine demand for conventional fossil fuels (Chapter 3). The electrification of almost everything and the Internet of Things will change the nature of the energy markets fundamentally.

Although the supply of fossil fuels may for all practical climate considerations be best regarded as infinite rather than about to peak, the demand may gradually leach away. Indeed, it has to – otherwise we face the possibility of dangerous climate change. The age of oil, and (eventually) the age of gas and coal, may not be over yet, but these two fundamentals will tend to undermine any future oil and fossil fuel commodity cycle. The oil industry, and the oil companies, may have been around for 150 years, but its best years are probably behind it, as we shall see in Chapter 10.

What all this means is profound: the price of oil may never go up so much again, except for short-term imbalances between supply and demand, and panics and peaks caused by war and political embargoes. There may be many of these, and with them lots of price volatility, but the underlying trend may be the opposite of what almost all our political leaders, company executives and investors have grown up believing. There may not be any more super-cycles, and the stable prices of the 100-year period to 1970 may really be 'normal' once again: not 'higher-for-ever' or 'lower-for-longer', but 'normal-for-ever'. We may have seen two great aberrations in the long-run trend, but the odds are against any more super-cycles.

There is a final twist to this argument. Imagine for a moment that medium-to-long-run prices may gradually be heading south. This means that oil today is worth more to a producer than oil tomorrow. It is no longer a good idea to slow depletion rates since the asset is of declining value over time. The result is that it is better to pump more today than wait for lower revenues tomorrow. What follows is self-fulfilling: the belief that fossil fuel prices are likely to fall triggers more supply which makes the prices fall.[20] It is the exact opposite of what dominated thinking at the end of the 1970s and up until the end of 2014. It has profound impacts on geopolitics and the position and role of the major oil producers and consumers. It fundamentally changes the game.

CHAPTER 2

Binding carbon constraints

If fossil fuel producers are challenged by the end of the commodity super-cycle and the resulting falling revenues, this pales into insignificance when compared with the existential threat to them posed by decarbonization. Put simply, the survival of the oil, gas and coal companies – and of the producer countries' revenues – is not compatible with mitigating climate change. This is the second predictable surprise.

Judging from the behaviours of both the companies and the producer countries, this existential threat has not sunk in. They appear to function in a parallel universe. The focus is still on finding and developing more reserves, and none has a credible post-fossil fuel strategy in place. Despite the non-governmental organization (NGO) campaigns, they don't see their assets as 'stranded', and they project rising demand through the coming decades.

To date the evidence has been largely on their side. In the face of rising emissions, global efforts to mitigate climate change have been largely inef-fectual. The Kyoto approach may have made things worse by encouraging trade diversion to China, and the Paris Agreement is little better. Then there is the key neglected factor: the price of oil. Most existing climate change policies have been built on the assumption of ever-higher fossil fuel prices, making the transition to low carbon much easier. Now the opposite has happened, with the double whammy of undermining the economics of nuclear and renewables, and increasing the demand for fossil fuels. The fall in oil prices is a big cut in the implicit price of carbon, and the new normal requires a correspondingly larger offset.

So are the oil companies and the producer countries right? Is the predictable surprise that, for all the noise, the end of fossil fuels is not nigh, and there are years of life left in them? Or is the predictable surprise that, notwithstanding the failures of past mitigation efforts, their days are numbered?

To answer these questions, and to see if the companies and countries are right, the starting point is to be ruthlessly realistic about the current approaches and to work through what lower fossil fuel prices mean for carbon emissions in the medium and longer terms. This sets up the business-as-usual scenario for the future of energy – a fossil fuel future. It is a world in which the policies follow the Paris pathway. Contrary to all the lobbyists' claims and vested interests, this would not be a world of 'stranded assets', but rather another relatively benign century for the oil and gas industry.

Fortunately for the planet, our energy future may not turn out like this. The more likely predictable surprise is that decarbonization happens, as the world turns to more promising policies and the technology is transformed. It will involve the jettisoning of many of the current approaches, and it will provide the second part of the double hit on oil – the revenues falling because of the price, and the demand falling because of decarbonization. The predictable surprise is that decarbonization will eventually kill the industry, and wreak havoc on OPEC, Russia and other producing countries. As we shall see, the picture for gas is rather different, and for coal it will have to be all about regulation and policy.

The conventional approach to climate change: why fossil fuels have prospered since 1990

In the parallel universe of many of the climate change campaigners, peak oil is just around the corner, if not already with us, and decarbonization is a win–win–win opportunity. As carbon emissions are reduced, new industries will be created, improving competitiveness and reducing consumer prices. The death of the fossil fuel industry is a small price to pay. What is there not to like about such a wonderful energy future, compared with the nightmare of ever-greater dependence on fossil fuel imports and ever-higher prices?

If it were true, oil companies would already be in steep decline, even as the oil price increased; investors would be getting out of these assets (even if the assumed higher price was making them more valuable). But sadly, it isn't: the conventional approach has not lowered global emissions; there are as yet no new significant current renewables companies in Europe at least (and several of the Chinese and US players have gone bust); and the impacts on competitiveness and prices have been negative, not positive. The oil companies have, as a result, barely noticed a pinprick on their asset values and profits because of climate change policies.

To see why, we need to start with the United Nations Framework Convention on Climate Change (UNFCCC) and the Kyoto Protocol that followed. Under Kyoto, developed countries agreed to impose caps on the *production* of emissions of a basket of greenhouse gases within their national boundaries. Developing countries would not be capped, though they would take 'measures'. Kyoto would be a stepping-stone, demonstrating that emissions could be brought down without serious economic costs, and gradually other polluters would come on board.

The Kyoto architecture was seriously flawed from the start. By reducing emissions in developed countries, the incentives to shift energy-intensive production to developing countries were increased.[1] Why produce steel in Europe and face the extra carbon-abatement costs instead of producing it in China? As long as steel production kept going up globally, shifting location at best made no difference to carbon emissions. In fact, it made the global emissions higher than they would have been without Kyoto since the Chinese electricity industry is very coal-intensive, and then there are the shipping emissions too. Emissions could fall in the developed, capped countries as they deindustrialized, while increasing in developing countries. That indeed is exactly what happened: there is no discernible impact on the growth of *global* emissions from the efforts made by Kyoto-capped countries. The recent stalling of this emissions growth path in 2015 is, tellingly, largely the result of China's stalled economic growth. Figure 2.1 paints the big picture confronting the advocates of 'more Kyoto'.

There were a number of further factors that made Kyoto worse than useless from a carbon-mitigation perspective – in addition to creating the illusion that something was being done. First, the US dropped out, to be subsequently joined at the exit by Canada and Japan. Second, the remaining countries – largely European – 'benefited' in emissions terms from the collapse of the Berlin Wall just before the baseline start (1990) and the

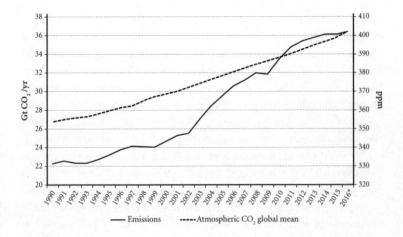

Figure 2.1 Global CO_2 emissions and atmospheric CO_2 concentration, 1990–2016

Notes: Emissions from fossil fuels and industry; Gt, gigatonne; ppm, parts per million; * 2016 represents forecast emissions.

Sources: CDIAC/GCP, 14 November 2016, http://www.globalcarbonproject.org/carbonbudget/16/ presentation.htm; Ed Dlugokencky and Pieter Tans, NOAA/ESRL, 2017, http://www.esrl.noaa.gov/ gmd/ccgg/trends/

consequent closure of much of the rust-bucket former Soviet industries after 1990. Even Russia could look good with this baseline.

The US under Bill Clinton had initially been very positive about Kyoto. Yet, as the negotiations unfolded into a formal Protocol, it began to distance itself. By the time it came to presenting the Protocol to the US Congress, it was obvious that few if any in the Senate would vote in favour, and hence it was dropped.

The Americans were not stupid in their approach to the Kyoto Protocol. If China (and India too) was outside, why would US business want to face what would, in effect, be a major trade distortion? Chinese companies could export their carbon-intensive products to the US without paying for the pollution, whereas Kyoto would have forced the US to impose costs on its domestic industries. Unless China played ball, there would be no obvious net gain in climate terms, and a significant competitive cost to US companies. Indeed, even President Obama realized this when he set off in late 2014 to negotiate a cap with the Chinese.

The American rationale was wasted on the Europeans. Europe's, and especially the European Commission's, position was based on its forecasts of ever-higher fossil fuel prices. It did not think there would be much of a competitive disadvantage to European companies because they would have 'cheap' renewables energy after an initial transition period. For European leaders, renewables were an industrial policy aimed at *improving* the European economy's competitiveness with the US and even China. It would take a transitionary decade or so, up to around 2020, and then the investments would pay off. If other countries failed to cut emissions, this would have the side-effect of boosting European competitiveness. They would lose out. Reducing carbon emissions would be, according to this logic, a sure bet.

In practice, it all turned out very differently. Europe's energy-intensive industries had a terrible time from the mid-2000s. These were assailed by cheaper Chinese imports, by the US shale gas developments, by the global economic crisis, and then by the Eurozone crisis. From the moment the Kyoto constraints were supposed to bite, for these *separate* reasons European emissions were not going to go up,[2] unless countries chose, from a climate perspective, really stupid options.

Sadly that is what Germany did, with the result that since 2008, despite the competitive disadvantages and the economic crisis, it has even managed on occasion to *increase* its carbon production, as we shall explore further in Chapter 8. Figure 2.2 reflects this. Germany was already on a downward path from 2000 (like most European countries). The economic crisis automatically sharply reduced emissions, but then, extraordinarily, they stopped falling, and at one stage actually went up.

The EU Emissions Trading System (EU ETS) made matters worse still, augmenting the renaissance of German coal. The Europeans prided themselves on translating the Kyoto caps into a total that would then be traded through a permits scheme. But, at the same time, they pursued a renewables programme *independent* of the EU ETS. The result was that as the renewables reduced carbon emissions it was easier to achieve the Kyoto caps as translated into the EU ETS. The carbon price was lower, and therefore the increase in renewables made more room under the EU ETS to burn more coal. One offset the other – almost perfectly. As the EU ETS carbon price fell, coal's new golden age began. Germany brought on stream more than 7.3 GW of new hard coal generation and 5.5 GW of brown (lignite) generation – nearly 13 GW in all between 2000 and 2015 – and it expanded

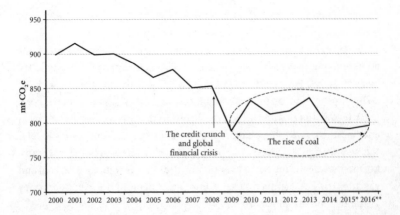

Figure 2.2 Germany CO_2 emissions, 2000–16

Notes: The numbers exclude land-use change and forestry; * 2015 is provisional, ** 2016 represents forecast emissions.

Source: Bundesministerium für Wirtschaft und Energie, *Energiedaten: Gesamtausgabe*, 5 May 2017, http://www.bmwi.de/DE/Themen/Energie/energiedaten.html

existing and opened new lignite mines.[3] As a result, it became one of the brownest rather than greenest of the major European countries. All of this was perfectly consistent with Kyoto. It was not until 2016 that Germany gradually started to reverse its dash for coal, but by then a great deal of damage had been done.

To be fair, the architects of Kyoto always envisaged it as a stepping-stone, and repeated Conferences of the Parties (COPs) were held to try to take the framework forward. Notable among these were Copenhagen in 2009 and Durban in 2011. The Paris COP in 2015 was the last under the Kyoto label as a new architecture is put in place.

Copenhagen in 2009 was widely regarded as a failure, which it unambiguously was. Large numbers of politicians and NGOs turned up for what was more a circus than a serious negotiation process. In the end, the US and China met *outside* the COP framework and, with a number of other countries, agreed the Copenhagen Accord, a series of 'pledges' for emissions targets. These were not legally binding, would not in any event add up to the emissions reductions necessary to meet the 2°C target, and left the US and China practically free to go on increasing emissions. The fossil fuel companies could (and did) relax: if this was all that the world leaders could throw at them, business-as-usual looked a pretty safe bet.

It got even better for them. At Durban in 2011 there was a serious attempt to put in place a post-Kyoto deal. The COP agreed that it would try by 2015 (at the Paris COP) to put in place a legally binding framework for 2030 targets.[4] But behind the ambition the reality was another failure. Durban was the point at which the Kyoto framework ceased to be global and became a European affair, as they were the only willing players left. Of the remaining developed countries, Japan had more pressing concerns: how to replace the output of over fifty nuclear power stations with coal and gas after the Fukushima disaster. Canada was bent on the expansion of its tar sands industry in Alberta.

Paris 2015: good politics, bad economics

The setbacks at Copenhagen and Durban did not stop the attempts at reaching a top-down global deal, and attention turned next to the Paris Conference at the end of 2015. Several things happened between Durban and Paris which encouraged world leaders to place their faith in the COP and global deals. These included the growing pressure to address the urban pollution in China from coal, the Obama–Xi Jinping deal, and the pressure of science.

As noted, the obvious flaw in Kyoto is that it does not properly take account of the sources of growing emissions in China, India and other developing countries. It was China's rapid economic growth, based on energy-intensive industries and coal, which drove up emissions after 1990, as well as causing the commodity super-cycle, just as emissions in the US and Europe were beginning to fall, as Figure 2.3 demonstrates.

Coal is dominated by China on an extraordinary scale. It moved from being a net exporter in the 1990s to being responsible for over half the total world coal trade in the 2010s.[5] An economy which doubles every seven to ten years, and has 80% coal in its electricity generation mix, is going to emit an enormous amount of carbon. The 80% is coming down, but a lower percentage of an economy which is growing this fast amounts to a colossal amount of extra emissions.

These emissions are produced in China, but they are not all consumed there. The export orientation of its economic growth means that carbon

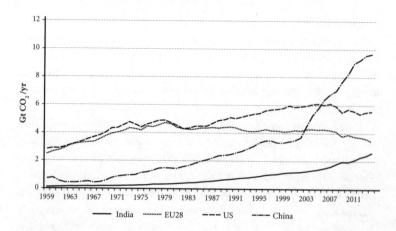

Figure 2.3 CO_2 emissions, 1959–2014

Source: CDIAC/GCP, 7 December 2015, http://www.globalcarbonproject.org/carbonbudget/15/presentation.htm

embedded in its exported production ends up being carbon consumption by the US and Europe, together constituting nearly half the total world economy. As noted, it makes little sense to curb carbon production in Europe or the US whilst allowing it to be embedded in China's exports. It is carbon consumption that matters in assessing the role of each country in emissions growth, and it is the absence of any consideration of carbon consumption under Kyoto, and the special circumstances of the European economy post-1990, that explains how Europe could reduce its own emissions whilst at the same time contributing to increasing global emissions.

Even in China itself the impacts of its coal-based economic growth have been becoming apparent. Key cities are blighted by the sorts of smog that the US and Europe experienced more than half a century ago in the 1950s. The air quality is so bad as to cause a large number of premature deaths and widespread health problems.[6] The Olympic Games in Beijing in 2008 illustrated the dangers. For the Chinese Communist Party, the risk is that protests might engender a return to the threat to its regime that it witnessed in Tiananmen Square in 1989. Something needs to be done for vital domestic political reasons, whatever the broader climate arguments.

China too has much to lose from climate change. It obviously wants to be viewed as a constructive global player on the world stage, and a gesture towards a new agreement which happens to fit with what it would do anyway has a lot of merit, as long as it does not derail the economic transformation on which the Communist Party relies for its political legitimacy.

Barack Obama had other objectives at Paris. By late 2015 he had entered the last leg of his presidency, and his legacy was one he was naturally focused on. No doubt he took the science seriously, but he also wanted to position the Republicans on the Luddite side of the climate debate. Donald Trump went on to adopt that position with relish. The windfall of shale gas had cut into the economics of the coal-fired power stations, so emissions were coming down irrespective of public policy, as shown in Figure 2.3. Moreover, Obama had learned the lessons of Clinton's failure with Kyoto: only if the Chinese accepted caps would Americans be willing to sign up too.

China duly obliged. But it was hardly the breakthrough Obama presented it as. China agreed to retain the right to carry on *increasing* its emissions for another decade and a half – to 2030 – after which it would cap further growth. This was neat politics on the part of the Chinese. Emissions would peak by then anyway, if not before, so it was a non-binding constraint. It showed China willing to play ball and be a responsible global citizen. In exchange, Obama agreed Europe-style cuts for the US, leaving it for Congress and the courts to vote him down on specific policy measures and take the blame – if the Republicans dared. It was crucial to Obama's strategy that there would be no legally binding regime which would mandate the need for ratification by Congress.

Meanwhile, the rest of the big and growing polluters remained on the sidelines. With a projected increase in the world's population of around 3 billion in this century, most of which will be in India, Africa and China, there will be a great demand for energy for even the most basic of human needs. For the fossil fuel industries it looks like a continuing boom. India stands out: it is overtaking China's growth rate, and has the potential to be the next great transition country. It too is very dependent on coal. Over five years to 2020, Coal India aims to double its production.[7] India's opening gambit for Paris was to avoid a cap, with the indication that emissions might peak by 2050. It suggested that it would need an implausible $2.5 trillion in compensation to act much quicker.[8]

Even less attention has been paid to Africa. Rapid population-growth countries like Nigeria and South Africa project major emissions growth. Then there are the Middle Eastern countries. So fast are their populations growing that in a number of countries the average age is below eighteen. To give a further flavour of the ballooning energy demands, Pakistan's population has grown from around 30 million at the time of partition from India in the 1940s to over 200 million today. It is building lots more coal power stations, financed largely by China. Indonesia, with a population of over 300 million, continues to create an annual smog covering Southeast Asia as its rainforests burn to make space for palm oil and other crops. It ranks as the fourth largest global polluter.

What these developments tell us is that there is a big gap between political aspirations and reality. As with Copenhagen, the pledges that the countries offered at Paris are not enough to meet the 2°C target. Indeed the emissions that would result if every country fulfilled its offers would exceed the 2°C by a wide margin. Worse still, there is no practical way in which the Paris signatories can be forced to deliver on their promises, even if the US had not subsequently opted out. If Paris-type agreements are necessary to halt emissions growth, then we are all in trouble. Our energy future will be a fossil fuel one, and the oil companies and producer countries are right to be relaxed about the impacts on them. They face no existential threat from the Paris Agreement. But this did not stop the negotiators deciding that although they were not able to agree on measures to achieve the 2°C target, nevertheless the target should be tightened to 1.5°C. As Richard Tol put it, 'In Paris, this near-impossible goal [2°C] was replaced by a more ambitious one.'[9]

False hopes for current renewables

If Kyoto and Paris are not going to halt the growth of emissions and hence the increased burning of fossil fuels, it is also rash to rely on the more bottom-up approach of promoting *current* renewables technologies, notably *current* wind, *current* biomass and even *current* solar. Not only are the costs high, the very arithmetic that would be required to add up enough wind turbines, agricultural land and rooftop solar panels at the global level to make much of a dent in emissions is missing.

There are two key features of wind and solar, the two main current renewables: they are *intermittent* and they are *low-density*. In the absence of large-scale storage, current renewables require back-up, and this is almost always from fossil fuels so far. Only fossil-fuel power stations can ramp up at scale fast. At the local level small diesel generators and small-scale gas peakers can also help. Although the lobbyists for these renewables are very keen to claim that each new wind farm or solar park powers X-thousand homes, and the media duly trot out these claims, the reality is that they do so for only part of the time. They both depend on the weather, and solar panels are useless at night. Yet at least solar is useless when demand is lowest. In theory, no wind means that the electricity system needs a complete alternative set of assets to meet demand: in other words, up to twice the level of the wind capacity. In practice there are ways of ameliorating the impacts. But for now the overall arithmetic is still awful and will remain so until the storage problem is solved, which it probably will be, but not yet, and not for some considerable time.

The intermittency substantially undermines the claim that these technologies are zero- or low-carbon. The energy that they produce is a blend of their outputs and burning fossil fuels, and this would be obvious if they had to compete on the same terms as conventional fossil fuel generation – by bidding *firm* (deliverable) power and having to contract for their own back-up, instead of others picking up the bill of these system costs they cause but do not pay. The costs of wind and current solar are not what their lobbyists typically claim, and they are not at 'grid parity' or close to being 'subsidy-free' on a full-cost basis, except in very special, and typically remote, locations.

The second problem is arguably worse: current solar and all wind are very low-density. The biggest wind turbines might reach 7 megawatts (MW), compared with 500–1000 MW for a modern coal- or gas-fired power station. The arithmetic follows: an enormous amount of land and shallow seas are needed to generate enough electricity for a modern economy, as well as the necessary back-up.[10] It just does not add up, and the conclusion drawn is stark: global warming cannot be seriously mitigated by current wind and current solar technologies. They have a role, especially in areas remote from major electricity networks and in sunny and windy locations. In some countries current solar potential is significant, yet even the recent advances in efficiency and the falls in costs come up against the sheer scale of electricity demand for countries like China and India, and in Africa and the Middle East. That noted, solar is better than wind and, as we shall see, its future may be *much* more promising. But for as long as the focus is on the current state of these technologies, fossil fuel companies can generally rest easy on this front, even if there might be serious impacts in specific markets, as long as there are no substantial improvements in their performances.

There are other ways of generating electricity without recourse to fossil fuels which do not suffer from either intermittency or low density. These are biomass, nuclear and hydro. Yet each has its own specific problems, which will limit their effectiveness over the coming decades.

Biomass promises to bypass fossil fuels by going straight from photo-synthesis to electricity and transport fuels via plants and trees. Large-scale biomass uses lots of wood and energy crops. In the electricity case, it is in essence the use of large log-burning stoves. Take the largest in the world – the DRAX power station in Yorkshire, England. In converting this massive coal-fired power station to biomass, a lot of electricity can be generated without intermittency. It is the largest importer of biomass wood from the US. The wood is converted to pellets, dried, trucked to the US coast, and then shipped across the Atlantic to the Humber Estuary, where it is loaded onto special trains, taken to a large purpose-built storage facility and kept cool, before being burnt as a low-density fuel to generate steam and hence electricity.

The argument that this sort of biomass produces sustainable energy rests on the claim that it uses waste wood that would not otherwise be used.[11] Put aside the environmental issues about the value to ecosystems of decaying wood and the specific case of DRAX, the general argument that biomass could be a significant part of the answer to climate change has to recognize that such material has other actual or potential uses, notably in the paper industry. As these biomass uses develop, the pulp and paper industry looks elsewhere for its products. Trees rather than waste wood are the obvious substitute. Trees and waste wood are ways of capturing and storing carbon. Planting more trees now, which over eighty years or so may recapture the emissions released from biomass burning now, is an offset. But this is disingenuous: the trees could be planted anyway, and emissions now may be worse than emissions in the future for reasons related to the new technologies that may be available later. Finally, like coal, burning wood results in other harmful emissions, notably particulates, and contributes to health risks.[12]

Advocates of this sort of biomass mostly realize this, and fall back on the claim that at least it is not as bad as coal (although even this depends on the relative efficiency of the power stations and the supply chain for the biomass).

However, if biomass from wood-burning has issues, these pale into insignificance compared with the consequences of converting corn into ethanol in the US, and rapeseed oil and wheat into biofuels in Europe. These measures are the product of a combination of trying to find alternatives to imported oil, and the lobbying power of farmers to gain greater subsidies. President George W. Bush's 2001 energy plan pandered to the Midwestern farmers in the US by mandating the use of biofuels in transport fuels and hence providing guaranteed markets.[13] The results have been awful, not least in driving up world grain prices and hence ultimately even forcing some communities to the edge of starvation.[14] It is argued that sugar cane for ethanol production in Brazil is better, but this relies on a questionable argument that it does not directly use former rainforest land. By taking up prime agricultural land, ethanol production has forced cattle ranchers to look for new lands, and these have in part come from cutting down and clearing rainforests, with all the consequences this brings for both biodiversity and carbon emissions. The Amazonian rainforest continues to be destroyed, year after year.

Biogas from animal and sewerage wastes can contribute to augmenting natural gas, primarily for heating, but it tends to be small-scale. Yet if current renewables technologies have made little impact on electricity generation and displacing fossil fuels in transport, they have arguably even less potential when it comes to decarbonizing heating. The easy bit is replacing coal and oil by gas, but replacing gas is altogether much more difficult, because heating is quite difficult to electrify. Solar can nevertheless contribute to heating water, and there is considerable scope for using solar gain in buildings. Electricity can address cooling.

Hydro relies on gravity, as indeed does tidal power. Many of the world's great rivers have already been dammed, but there are plenty more opportunities to do so. Having constructed the Three Gorges Dam on the Yangtze, China now plans very major dams on the Mekong. And there are lots of opportunities for the Amazon, the Congo and other great rivers. But all of these come with enormous environmental costs, and it is hard to argue that the effects on biodiversity and the other damage are worth the carbon saved, given other options. Tidal power and tidal lagoons have enormous potential in very specific locations, but also come with high costs.

Nuclear power offers a much brighter prospect, at least in theory. It can generate very large amounts of electricity, and it is baseload. There is no obvious shortage of uranium. It is possible to envisage large-scale nuclear construction that would make serious inroads into electricity supply. Indeed, it has done so in the past, in the US and in much of Europe, in Japan, and now in China and India too. Yet for a variety of reasons nuclear's overall global prospects are probably going backwards rather than forwards for at least the next couple of decades.

The Germans, much of Western Europe and Japan have all turned their backs on nuclear (although Japan may very gradually and partially revert). Even France has legislated to reduce the share of nuclear from 80% to 50% of its electricity supply. With most of Japan's nuclear power plants idle, and with the first generation of nuclear plants in the US and Europe coming to the ends of their lives, nuclear will in aggregate probably go into decline before making a net positive contribution to reducing emissions.

With a vulnerability to public opinion, the costs of nuclear tend to get asymmetrically ratcheted up. Every time there is an accident, work on new plants stops, there are safety reviews and then the standards are tightened. Building a new nuclear plant in Europe is now extraordinarily expensive. In Britain the projected cost of the 3.2 GW plant at Hinkley Point, including the financial costs, is around £22–£24 billion. Even if it is built on time and on budget (something that has not been achieved in the two PWRs of this design in Finland and France), it will still be the most expensive power project in the world.

Elsewhere costs are much more reasonable and build-times shorter. China stands out in this regard, although the French-designed PWR at Taishan is already late. But why are these Chinese nuclear plants apparently cheaper? State secrecy makes it impossible to know whether the safety standards are weaker and conditions poorer for the construction workers. The fear is that the fast-track, secret programme is vulnerable to an accident. The recent industrial explosions at Tianjin in September 2015 (173 reported dead) and the vast landslide of construction waste at Shenzhen in December 2015 (91 dead) reflect a very different approach to safety in an economy riddled not only with secrecy but also corruption.[15] It would be surprising if the Chinese nuclear industry is exempt from these failings.

The oil companies' view of the future

For these reasons, nuclear is not going to come to the rescue anytime soon. With biofuels and biomass making only limited inroads into the transport and

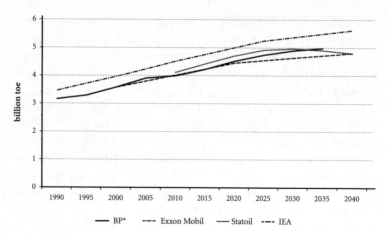

Figure 2.4 Oil consumption projections

Note: * Energy consumption for BP liquids includes oil, gas-to-liquids and coal-to-liquids.

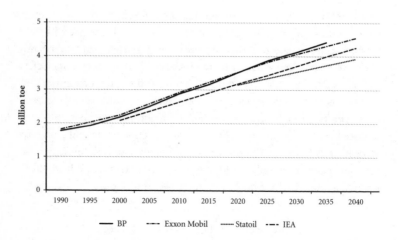

Figure 2.5 Gas consumption projections

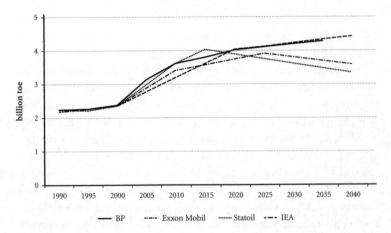

Figure 2.6 Coal consumption projections

Note: IEA figures represent New Policies scenario, Statoil figures represent Reform scenario.

Sources: BP, 'Energy Outlook 2035', February 2016, http://www.bp.com/energyoutlook; Exxon Mobil, 'The Outlook for Energy: A View to 2040', 2016, http://corporate.exxonmobil.com/en/energy/energy-outlook; Statoil, 'Energy Perspectives', 2016, http://www.statoil.com/en/NewsAndMedia/News/EnergyPerspectives; IEA, *World Energy Outlook 2015*, http://www.worldenergyoutlook.org/, © OECD/International Energy Agency, *World Energy Outlook 2015*, IEA Publishing

electricity markets, and with current renewables not able to provide electricity at the required scale, cost or baseload, the fossil fuel industries might be quite rational in seeing a bright future for themselves for several decades. This indeed is pretty much what the main oil company projections show out to 2040 and even 2050 – for oil, gas and coal. Figures 2.4, 2.5 and 2.6 show that, for oil and gas, they project demand to keep going up for the next couple of decades at least. Coal is less clear-cut, depending on policies.

Inter-fuel switching

Although the current top-down Kyoto and Paris approach to a global set of caps and the bottom-up approach of current renewables, biomass, nuclear and hydro do not fill the carbon gap, it does not follow that political leaders are going to abandon attempts to mitigate carbon emissions. For the science will not go away, and the evidence keeps mounting up. Having tried all the other alternatives, the oil companies should recognize that governments might finally do the right things. That means starting to get out of coal and switching from oil – in both cases initially to gas.

Whilst oil companies may have so far avoided serious damage to their business models, the same cannot be said for coal. Not all fossil fuels are in the same boat, and the switch *between* the fossil fuels is already underway. Our energy future may involve lots of fossil fuels, but the mix may change significantly. This switching is about relative costs and relative pollution, and hence it is largely immune to the general fall in fossil fuel prices discussed in the previous chapter.[16]

A great deal of political capital has been used up making the case for current-generation solar, current wind generation, and biomass and biofuels, but the immediate villain is coal. Coal is the dirtiest in carbon emissions, as shown in Figure 2.7. It produces not just CO_2 but other green-house gases too, and has very low thermal efficiency once the full cycle – from the opening of mines (and the methane they leak), through the

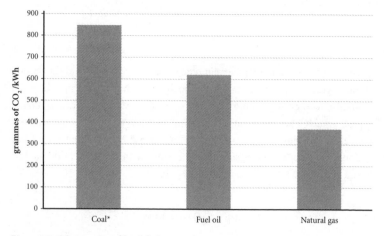

Figure 2.7 CO_2 intensity of fossil fuels

Notes: * Average of anthracite, coking coal, other bituminous coal, sub-bituminous coal, and lignite; KWh, kilowatt-hour.

Source: IEA, 'CO_2 Emissions from Fuel Combustion: Highlights', 2011 edition, http://www.iea.org/media/statistics/co2highlights.pdf, © OECD/International Energy Agency, 'CO_2 Emissions from Fuel Combustion: Highlights', 2011, IEA Publishing

transport of the bulky coal, to the combustion processes and the waste, and finally the losses in the electricity networks – is taken into account. Coal-burn yields smog and particulates, causing urban pollution and contributing to premature deaths. As has already been noted, its share in world energy has crept up from around 25% to almost 30% since 1990, and it has caused most of the emissions growth. But its days may be numbered – or at least checked. It is among the cheapest fuels to displace. Switching from coal to gas reduces emissions fast. Gas has roughly half the carbon emissions of coal, and this is before taking into account all the other pollution that comes from mining and burning coal. Gas is the cleanest of the dirty fuels. The future is therefore *relatively* better for gas in a gradually decarbonizing world – a fact not lost on some of the major oil and gas companies, most notably Shell. In contrast to the German example, the US shale revolution has shown how this can be done. Accordingly, the results for US growth of carbon emissions have been at least as good as, if not better than, those of Europe since the mid-2000s, especially when the relative economic growth rates are taken into account.

The wiser climate change campaigners have finally grasped this point, and one of the most notable features of the public debates in the run-up to the Paris Conference was the attack on coal. Even Germany has recognized that coal will have to go if it is to meet its green aspirations. Britain has publicly committed to phase coal out by 2025, and many countries already require CCS for any new coal plant, and hence price new coal stations out of the market. Conveniently, national and regional action can do the job. No Paris-type approach is needed.

These campaigns have been reinforced by a further significant development. The big oil and gas companies in Europe have publicly turned their guns on coal (though they were not immediately supported by Exxon and Chevron in the US). These companies have an opportunity to be seen to be taking a responsible approach to climate change and to get on the front foot, and have not only pushed the case for gas against coal, but have also supported the case for carbon pricing, which would further aid this switch since it would bear down disproportionately on coal. The most notable convert to this position is Shell, but Statoil, Total and Engie have now joined in.

Elsewhere, coal's future has also come under pressure. China's pollution problems have put pressure on coal plants in the east near the big cities. Its

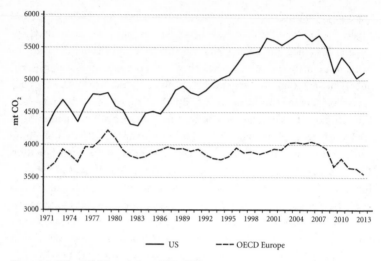

Figure 2.8 US vs EU CO$_2$ emissions, 1971–2013

Note: OECD, Organisation for Economic Co-operation and Development.

Source: IEA, 'World CO$_2$ Emissions from Fuel Combustion', OECD/IEA, Paris, 2015, http://wds.iea.org/wds/pdf/Worldco2_Documentation.pdf, © OECD/International Energy Agency, 'World CO$_2$ Emissions from Fuel Combustion', 2015, IEA Publishing

energy future is no longer so overwhelmingly based on coal. But not everyone can opt for gas, for the good reason that not every country has either gas deposits or the liquefied natural gas (LNG) and pipeline infrastructures to import it. Yet this is coming: LNG is gradually becoming a global business, and the US switch to exporting gas has aided this process (both through direct exports, but more significantly also by freeing up what it would otherwise have been importing), as have the massive gas projects in Australia and Qatar. Conventional pipelines are being built across the world – for example, from Russia to China, from Russia to Europe, and from Central Asia to China. The development of new shale gas opportunities in countries as diverse as Argentina, China and Algeria will add to global supplies.

The global age of gas will be aided by climate change concerns. It would happen anyway, but the scale will be enhanced as coal is squeezed out of the energy mix. In due course, gas will probably squeeze out oil too, as transport is converted to gas directly and to electricity, which will in turn be partially generated from gas for a long time to come. This is for the next decade rather than immediately, but as we shall see in the next chapter, the impact of gas and then electricity on transport's demand for oil will make a nonsense of all those crude company projections set out above, which look at income growth, translate it into the growth in car numbers, and then read across to greater demand for oil. Transport demand can go up without oil demand going up. Not yet perhaps, but sooner than many in the oil industry realize.

Carbon prices and regulation

Although fuel-switching will happen, the grip on the energy future of fossil fuels as a whole will be tightened by the impact of the drop in oil, gas and coal prices. When fossil fuel prices tumbled, it was equivalent to a sharp cut in the price of carbon for those countries that have a price, and an increase in the implicit subsidy to carbon pollution for those that do not.[17] It is easy to assume that the lower carbon price this implies means higher carbon emissions. Yet it also assumes that governments will not react to the falling fossil fuel prices and use the opportunity to raise carbon prices. Following on from the anti-coal policies being pursued in the US and Europe through regulation, they probably will raise the price of carbon, and indeed they have already started, though it may be via general energy taxes as well as specific carbon taxes. The price of energy to customers has rarely been set by the world market prices, and these typically have little relationship with the underlying costs of production. Energy prices are driven by royalties, taxes and in some cases subsidies. The marginal cost of production in the Middle East may be as low as $5 a barrel. Even at the low point in January 2016 the market price was $27 a barrel. The price at the fuel pump in Europe is an order of magnitude greater.

The obvious policy response to falling fossil fuel prices is to increase the carbon price. This can be done in two ways: by emissions trading (such as the EU ETS); or a carbon tax. Politics favours the former; economics the latter.[18] Emissions trading requires a carbon emissions cap to be created, then divided into individual permits which can either be given away or sold by governments to polluters, and then traded to create a market price, provided that there is no interference with the quantity set in the cap. The great advantage is the certainty that the cap will be met (or at least for those emissions sources it covers); the great disadvantage is that the cost is left open.

Polluters obviously prefer the permits to be given to them – or 'grandfa-thered'. This way the money is kept by them and not transferred to govern-ments, as it would be if the permits were auctioned or if there were a tax. This is what they initially got with the EU ETS. Even better for the polluters, as the economic crisis and deindustrialization took their toll on the European economies, the price ended up very low, so low indeed as to have no impact on anything except the profits of the traders in the permits.

The EU ETS has produced a short-term, volatile and low price, when what is needed is a long-term, stable but rising price.[19] This is what a carbon tax could achieve: it can be set so as to recognize that the existing capital stock is given (and hence the emissions in the very short term are hard to change), but to signal in the medium term switching from high-carbon technologies (like coal) to lower ones (like gas), and in the long term to signal incentives to develop new low-carbon technologies.

Given the power of industrial lobbying, including by the oil and gas companies, it is no surprise that the Paris negotiations focused on the intro-duction of emissions trading schemes in developed countries, and that they have sprung up in China, the US and elsewhere. But the case for a carbon tax is not lost. On the contrary, almost all developed countries already tax energy, and these energy taxes, especially for transport fuels, are proxies for carbon taxes. Everyone interferes with the price of energy through taxes. These energy taxes are a third, implicit, way of taxing carbon. A number of countries already have their own carbon taxes, and some, such as Britain, have floor carbon prices.

Now that the need to switch from coal to gas is more widely recognized, the pressure is on to find ways of increasing the carbon price. The EU is currently committed to doctoring the EU ETS to push its price up, though this is clearly in direct contradiction to the idea of a permits system: fix the number of permits, and let the price adjust. By back-loading the permits and moving on to removing them from the market, it can target a desired price. It is, in effect, an incredibly inefficient way of ending up with a carbon tax: setting a target price, and then manipulating the quantity to achieve it.[20]

There is an additional pressure that has little to do with climate change. Governments need money. After the financial crisis and the massive increase in sovereign debt, and the continuing problem of deficits, the need for tax revenue is pressing on many countries, from China and Japan through to European countries and the US. Some countries have already reached the economically efficient and politically acceptable levels of income tax, and consumption taxes are mostly high. Governments need new tax bases, and carbon offers an opportunity to cloak a revenue-raising mechanism in the guise of an environmental necessity. This pressure can only grow.

Further ahead: the stranded asset debate

In the short term, it is the coal-to-gas switch that can have the greatest impact on carbon emissions, and this in turn would herald a big change in the nature of fossil fuel markets. Given the much wider distribution of gas resources, the geopolitics of gas differs from that of oil (and coal). This will have a negative impact on the power of OPEC, as we shall see in Chapter 5.

The climate change problem will not, however, be *solved* by gas. Gas is a temporary and urgently needed expedient in our energy future. In time, gas will have to be phased out alongside oil and coal. The best way to think about the impact of climate change policies on the fossil fuel markets is the slow but persistent scientific pressure building up, carbon prices (whether explicit or implicit) and regulation cutting into revenues and demand, and the very gradual downward pressure on production. The drop in fossil fuel prices makes this harder, but it does not stop its trajectory.

Academics have worked out how much more fossil fuels can be burnt whilst meeting the 2°C target.[21] They regard this as a total envelope, beyond which fossil fuel reserves will have to be left in the ground. These will become, it is claimed, stranded assets. Oil companies are attacked for developing new marginal resources, be they tar sands or Arctic drilling. Coal companies are confronted with head-on disinvestment campaigns. Shareholders' meetings are bombarded with demands that carbon policy be addressed.

These calculations, and the campaigns behind them, have the great merit of identifying the inconsistency between the 2°C target (or even the new Paris 1.5°C target) and what is actually going on in energy markets, what investors believe and what the companies are doing. These do not add up. They are not part of the fossil fuel industry's parallel universe, or their projections, as we have seen. Yet they do illustrate that achieving a gradual phase-out of fossil fuels, and at the same time providing enough energy to underpin economic growth and the population increases to come, are inconsistent without a major shift in technologies. Existing technologies, with the possible theoretical exception of nuclear, cannot bridge the gap. Willing the end requires the willing of the means. These do not yet exist.

Some are seduced by the ultimate get-out-of-jail-free card. If only the carbon could be put back in the ground (or under the oceans) then the fossil fuel industry might be carbon-neutral. Predictions of its demise would then be premature. The concept is simple; the practice is anything but. Carbon can be captured (at considerable cost). It can be piped into underground stores. These stores can be sealed. But there are big obstacles. In addition to the costs, there is the absence of enough holes and sinks. Where exactly is all this carbon, on such a vast scale, to go? CO_2 requires much more space than the coal, oil and gas it is produced from, by an order of magnitude.[22]

Even if CCS was to play a part, as it surely will, it does not follow that it will come to the rescue of the fossil fuel industry. There can be little doubt that we are going to overshoot on emissions, and that therefore at some stage we will need negative-emission technologies to manage down the stock of carbon in the atmosphere. Agriculture and land use will add more to the energy sector's emissions to feed the growing population.

The CCS option is needed *as well as*, not instead of, reducing the burning of fossil fuels. There is no way out of the carbon constraint, useful though this technology will undoubtedly be. Planting trees is no substitute for burning them in biomass power stations. CCS is no substitute for more oil, gas and coal combustion.

It is therefore quite rational for the oil, gas and coal companies to go slow on CCS. If it were the solution, if it were to give them an indefinite life in a decarbonizing world, then they would surely be pressing ahead. But they are not. The pilots and demonstration projects are trivial compared to the scale of the problem, and most governments have backed off from any early enthusiasm they harboured. The companies fall back on their assumed predictable surprise: they assume that nothing much will be done about climate change, and certainly nothing on a scale to head off the rising demands for oil and gas, even if coal falls off its dominating perch.

But before we fall into line with their preferred energy future, another, third, predictable surprise should not be dismissed. Indeed it is likely to come to pass. For the changes in the climate will not go away. Scientists tell us they will only get worse, so the pressure to change can only get stronger.

The implication is both obvious and missed by both sides of the debate about stranded assets and disinvestment. What is required are new technologies, not simply lots more current renewables, from which the industries have little to fear at the global level. If the fossil fuels are to be left in the ground, something has to replace them. Investors and companies are right to be sceptical about the claims of campaigners that the assets are actually going to be stranded on the basis of current policies and the evident failures of the Paris negotiations to properly address the full suite of developing countries, including China and India. But they are wrong if they think that the only challenge to their assets is current renewables technologies. Ahead lies a cornucopia of new technologies, which have every chance of transforming the energy landscape in the next couple of decades. These new technologies have the potential to be at least as radical in their impacts as the coming of the internal combustion engine and the coal power stations at the end of the nineteenth century. The age of coal, oil and even gas may be gradually drawing to a close, as may be the days of some of the incumbent companies.

CHAPTER 3

An electric future

The age of fossil fuels will not come to an end because we run out of oil, gas or coal, and current carbon policies are not going to scare fossil fuel company directors. But what is a very real threat are the new technologies, and possibly sooner than commonly assumed. This is the third predictable surprise.

After the plethora of great discoveries towards the end of the nineteenth century and in the early twentieth century, the energy world was one of comparatively little change for the rest of the twentieth century. After the three great inventions – coal power stations to generate electricity, the internal combustion engine and the associated oil refinery technologies, and petro-chemicals – there were few subsequent big breakthroughs. The main additions were nuclear and gas turbine power stations, both mid-century. Even in the latter case, combined-cycle gas turbines (CCGTs) could not be legally used to generate electricity in the US and Europe until 1990.

And that is about it. Some nineteenth-century electricity cables are still in use in London, and transmission technology has not delivered the sorts of 'superconductivity' once promoted. Pipelines remain just pipelines. Tankers are bigger but still just tankers, though LNG has been added. Go back a century and compare a 1970s-vintage coal power station with its early progenitors in the nineteenth century and most of the modern plants could be readily understood then. New super-critical thermal plants have very recently advanced coal generation considerably, but the basic concepts remain the same. It is also true that computers and IT have changed control rooms beyond recognition. But that is about it. This continuity is also reflected in refining and the internal combustion engine. Incremental changes? Yes, lots of them. Fundamental breakthroughs? No.

As a result, the sorts of industrial structures witnessed today in the oil, gas and electricity industries remain the solutions to the cost structures that these traditional technologies reflected. In oil and now gas, large vertically integrated international companies sprung up to carry the high fixed and sunk capital costs of the drilling rigs, platforms, pipelines, refineries and associated infrastructure, and they built large portfolios to spread the risks. In electricity, vertical integration dominated from the outset – typically the full chain from the fuels to the final customers' meters. The absence of storage, and passive demand, put a premium on centralized control. The sunk and fixed costs made monopoly essential to spread the risks and recover upstream costs. The oil and electricity companies tended to create monopoly power through dominating supply from the wellhead through to the filling station, whilst in electricity governments tended to own the companies at the local and then the national level (in Europe), and imposed statutory monopoly to make absolutely certain customers had nowhere else to go (in the US and Europe).

After so much stability for so long, and such dominant technologies, many politicians and executives take it as given that these are the structures that will continue to dominate energy production and supply this century. They find the very idea of rapid technological change hard to grasp. They see the future in terms of Shell, BP, Exxon and Chevron, joined at the table by the great national oil companies of the Middle East and China. They see RWE, E.ON, EDF, GDF (now Engie), Exelon, Duke Energy, Southern Company and AES Corporation in a similar way. It was once assumed that the great telecommunications monopolies (like AT&T) and the great computer companies (like IBM) were the future in their industries.

Complacency about technological change, combined with belief in higher prices and the return of the super-cycle, as well as cynicism about the prospects for decarbonization, pervade the energy sector, and lead the main actors to project onto the future a past with which they have long been familiar. They drive the car forward by looking out the back window. Energy strategies and policies focus on solving twentieth-century problems created by long-established technologies.

All this is about to change as the ground slips beneath their feet. There is now a large and bewildering number of new technologies emerging across the energy sector. These are largely focused on electricity, and it is electricity that is making inroads into almost all energy markets, from cars to support for new manufacturing technologies. This chapter is about the scale of this technological change over the medium-to-longer term, and why, over this time horizon, it may mean that oil and (eventually) gas prices go down rather than up, and why coal, oil and eventually gas may be left in the ground. New technologies, following on from the end of the commodity super-cycle, and incentivized by decarbonization, are transforming the future of energy.

The future is electric

From a slow start back in the 1970s, electricity is increasingly the way in which energy is used, as the chart below illustrates. With the coming of the

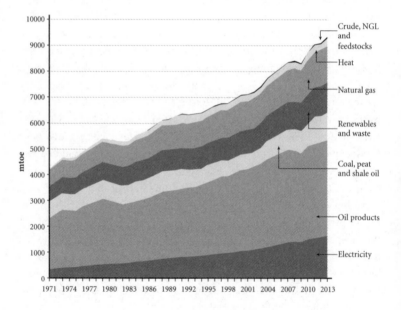

Figure 3.1 World final energy consumption by source

Note: NGL, natural gas liquids.

Source: IEA, 'IEA Headline Global Energy Data', 2015, http://www.iea.org/statistics/

new information technologies from the 1980s, electricity has been gradually gaining ground across the board, with only the transport sectors and some heavy industry holding out.

The ground that has gradually been gained by electricity has been a function of the more general transformation that the new communications technologies kicked off, and in particular digitalization. So far the impact has been small, but it is about to accelerate. Almost any object and most services can be digitalized, and once this happens the resulting data becomes the critical asset. The handling of this data is almost always powered by electricity. It is what makes computers tick. Data cannot be oil-, gas- or coal-fired.

The relationship between digitalization and electricity demand is multifaceted. Computers have directly increased the demand for electricity. The requirements of modern communications technologies have pushed electricity up the energy league tables. But this is only the beginning: it is what happens to the digitalized information that really transforms the role of electricity. Everything that is digital is electric, and as everything is digitalized, almost everything becomes electric.

This trend can be seen in cities and their information infrastructure, in buildings, offices and factories. Robots, supported by AI and additive manufacturing (3D printing), to which we return below – all digital technologies – are key components of the future of manufacturing. Transport is slowly going the same way. Railways continue to be electrified.

Perhaps a better way to talk about the trend towards electricity is to focus on those areas that are proving resistant. They include aviation, some aspects of heating, and some heavy industrial processes. But remarkable as it might seem, even in these hard-to-electrify areas, electricity is almost always at least a possibility.

So in thinking about the energy futures, and in particular the future of fossil fuels, we should think electric. The conventional fossil fuels are gradually being forced out of their main markets, and in time they will be forced out of even their most secure markets. Their future is increasingly bound up with the role they play in generating electricity and how competing ways of generating that electricity will gradually squeeze them out. It is in electricity that the major technological progress is being made, across the full vertical industry chain.

New electricity-generation technologies

For most of the twentieth century electricity was generated by burning coal. It contributed around 80% of total generation in the US, Europe, Japan and China until the 1970s, supplemented by nuclear thereafter. France and Japan switched significantly in the nuclear direction, but for most of the rest the electricity fuel mix was roughly 80:20 coal/nuclear. Oil played a role where it was abundant and cheap, notably in the Middle East, but globally the world of electricity relied overwhelmingly on coal. It still does. It makes up over 40% of electricity generation.

There have been many advantages to relying on coal. It is widely distributed and can be stockpiled so there are few if any security-of-supply problems. Its abundance makes it cheap. It varies in composition, but the robustness of coal power stations can cope with a lot of variance and still survive for decades – in many cases more than half a century.

But as described in Chapter 2, these workhorses have one huge disadvantage. They are horrendously polluting. The search is therefore on for alternative ways to generate electricity. The good news is that there are lots of them. Low-carbon generation comes in a number of guises – solar, wind, geothermal, nuclear and gravity. Of these, gravity in the form of hydro is limited and has serious environmental impacts; tidal is typically expensive and more long-term; and geothermal has potential as both a large-scale heat source and a local one (and hence probably has a bigger role for heating rather than electricity generation). Although any and all may surprise, for the medium term, this leaves nuclear and solar. (Wind has little to contribute at scale, as explained in Chapter 2.)

Nuclear has always had great potential. It can generate an awful lot of electricity. It is secure in the sense that it does not depend on distant resource-cursed suppliers and the likes of Russia and the Middle East. The technology has been around for more than sixty years, and a lot of reactors (over 400) are working worldwide. It would be perfectly possible to generate much of the current and future electricity demand from nuclear.

And yet we don't – and we probably won't without a technological breakthrough. Recall the problems raised in the previous chapter. Nuclear has three key drawbacks: it is dangerous (and certainly widely perceived as such); it creates nasty waste; and it is expensive. Danger comes from the consequences of accidents, the link to nuclear weapons and, more recently, the perceived vulnerability to terror attacks. The major nuclear accidents involving Three Mile Island (1979), Chernobyl (1986) and Fukushima (2011) are well documented, but there have possibly been others which were kept off the radar. Secrecy is pervasive in countries like Russia and China, and until modern detection equipment became available it is possible that much went unreported. The British tried and failed to keep the 1957 Windscale disaster secret. The major accidents come around every decade, and they show that even advanced countries cannot guarantee safety. Moreover, more diffuse dangers are associated with the disposal of waste materials.[1]

None of these incidents has, however, been anything like as bad as the deaths, injuries and impaired life expectancy related to coal mining. But that is not the point: nuclear accidents *could* be devastating, and they scare people. No recent coal-mining disaster or coal-related health scares have attracted the media coverage given to Fukushima, and there are many who would prefer not to bear this risk. Nuclear proliferation for military purposes is practically impossible without civil nuclear facilities. Waste hangs around for a long time and very few countries have yet come up with permanent solutions. Finally, few countries have ever made nuclear power cost-competitive, not least because of the safety concerns and the associated regulation.

The conclusion that follows is that the *current* nuclear technologies, notably the large-scale PWRs, boiling water reactors (BWRs), *reaktor bolshoy moshchnosty kanalny* (RBMK) and advanced gas-cooled reactors (AGRs) in the US, Japan, Russia and China, are likely to remain out of the market and, for many, out of the politics too.[2] This will not stop some of them being built, mainly in China where safety standards may be lower. But the new build will probably not keep pace with the global closures of existing stations through old age or political interventions.

But what of next-generation nuclear? Here the possibilities are enormous and varied – fast breeder reactors (FBRs), small molecular nuclear reactors, and new fuels such as thorium – but the horizons are very long. Then there is the great hope of the nuclear advocates: the perennial promise of fusion – the sort of nuclear reaction that powers the sun.[3] The problem is the gap between these concepts and the science on the one hand, and the reality of bringing one of these technologies to market on the other. None will make any serious impact until at least 2050, and therefore the fossil fuel companies can rest easy about the threat of a nuclear breakthrough driving them out of business for decades to come. Our energy future will not be dominated by nuclear for quite a while, not until at least the second half of the century, if at all.

Even the various sorts of small modular reactors (SMRs) which, in theory, could be manufactured in large numbers and deployed widely, do not look like making much difference in this half-century. These are typically scaled-down versions of large reactors, like the PWR. They would suffer inefficiencies from the loss of scale effects, and these would have to be offset by gains from manufacturing lots of them. But even if this trade-off works in their favour, there is still the public reaction to take into account. It is hard to imagine, for example, that the citizens (and voters) in numerous locations are going to be any more welcoming than those in a small number of places where large PWRs are planned. The large reactors tend to bring lots of local jobs and have an economic upside for local people. They tend to be in or near places where there is a considerable nuclear history. But imagine the reaction in, say, ten or twenty major towns and cities confronted with their own 100 MW nuclear reactor on their doorsteps. More promising perhaps, SMRs can also be transportable, notably on reactor ships, and these are already widely deployed. But the key point here is in the name – they are small.

If nuclear is not on the cards as a major source of electricity generation in the medium term, we are left with solar among the low-carbon options. Solar has a lot going for it and its bright prospects are effectively infinite. The sun's energy is so abundant that more energy is transferred to our planet in an hour's sunlight than the entire global electricity industry can generate in a year. The sun's nuclear reactor has produced lots of oil, coal and gas over hundreds of millions of years through photosynthesis, and it continues to produce vegetation, sequestering carbon, and will do so effectively for ever. The heat from solar radiation creates the winds too. We have been harvesting these indirect forms of solar energy for all of human history.

What has changed is the ability to convert sunlight *directly* into electricity, bypassing the more inefficient route via photosynthesis. Photosynthesis will still play a part in the energy mix, however, through biomass and biofuels. The plant material is converted into fuels on a large scale, notably in the US (corn) and Brazil (sugar cane). It is also, as we noted in the previous chapter, burnt directly in power stations. But none of these adds up to sufficient scale to meet future energy demand, and all ultimately are hampered by the sheer inefficiency of photosynthesis. We are therefore left with direct solar capture.

To date, the process of direct capture has also been fairly small-scale and inefficient. It is true that very considerable progress has been made in reducing the cost of solar panels, and that solar parks have increased the scale. 'Swanson's Law' claims that the price of solar photovoltaics drops 20% for every doubling of cumulative shipped volume.

This cost improvement looks impressive when set against the very inefficient and costly starting line. Figure 3.2 shows just how much progress has already been made. Current solar is therefore much better and cheaper even than a year or two ago. Extrapolating Swanson's Law would take it much further, but caution is in order: the 20% is an empirical observation of progress so far, and cannot be predictive. In any event, the result of this progress is that much of the solar panel investment in the last ten to fifteen years reflects the inefficiency in earlier versions. Germany has so far spent over €200 billion in subsidies on this lower-efficiency vintage of solar panels, and overall renewables subsidies amount to €20 billion per annum.[4]

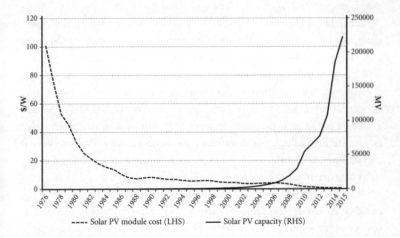

Figure 3.2 Solar costs and capacity

Sources: Swanson, R., 'A Vision for Crystalline Silicon Photovoltaics', *Progress in Solar Photovoltaics*, 14(5), August 2006, pp. 443–453; REN21 Steering Committee, 'Renewables 2016: Global Status Report', Paris, http://www.ren21.net/wp-content/uploads/2016/06/GSR_2016_Full_Report_REN21.pdf

Whilst, as noted, it is true that the cost of manufacturing these conventional solar panels has fallen, and that this is due in part to improvements in manufacturing techniques and scale effects, it is also the case that the Chinese companies, which have captured so much of this market, have been subsidizing and even dumping the products on their markets. Several US solar companies have gone bust partly as a result, notably Solyndra in 2011 and SunEdison in 2016.

What is needed for solar to really cut into the demand for fossil fuels is to address the two weaknesses in current technology: only a small part of the light spectrum is harvested; and the materials in the solar panels are still relatively primitive. Solar cells harvest the visual light spectrum to generate electricity. They cannot yet harvest the infrared and ultraviolet sunlight, which comprise, respectively, 47% and 46% of the light spectrum. The great challenges facing solar generation follow: how to open up the light spectrum from the current 7%; and how to harvest the energy through more efficient materials and mechanisms. Both are 'science' problems; neither is primarily about learning-by-doing. In other words, deploying lots more *current* solar panels won't necessarily help. Yet it is the latter that has taken such a huge share of public subsidies, with much less left for research and development (R&D).[5]

Breakthroughs are nevertheless being made. New parts of the light spectrum are being accessed for electricity generation, expanding the solar potential by orders of magnitude. New semiconductors made from materials such as indium, gallium and nitrogen are helping to develop new applications. The infrared in particular is being opened up. Solar film, perovskite layers on top of silicon,[6] and the use of nanotechnologies and graphene are all contributing to a solar energy future. Nothing comparable appears to be on the horizon for wind and wind turbines. Progress here is incremental, and ultimately limited. Wind will play a role, but it is unlikely to be decisive in decarbonization.

It is now possible to at least *imagine* a world in which almost everything could be coated in solar film, and where much more of the light spectrum is opened up for capture. This could be a revolution in the making. The energy source is free, and the marginal costs are zero, or close to zero. Imagine if all windows, buildings, cars and even clothes were coated in solar film. This would be a massive extra supply of electricity, and it would be largely decentralized and local, as opposed to the large-scale centralized coal and nuclear systems of the twentieth century. There might even be large-scale solar power stations, concentrating solar to generate steam too.[7] It might represent a source of electricity that is cheaper than the fossil fuels. It would then no longer be feasible for large coal mines, and ever-bigger coal power stations (even if they were modern super-critical ones), to compete in this solar world. This would be a world in which almost everything was electric, and most electricity was solar.

This is not the place to adjudicate on the science or the applications. That is for the technical experts. The point here is to identify some of the broad *possibilities*, and to consider whether these show any revolutionary promise. It is also to think through the practicalities. If the domain for improvements is known, so too is the relative absence of risk. The neat thing about solar is that it carries none of the risks identified with our other main low-carbon option, nuclear, whilst at the same time having the *potential* of scale. It also comes in bite-size bits, not large-scale nuclear units. Given the doubts about CCS highlighted in the previous chapter, and the limitations on more hydro and thermal sources, it is the main game in town. Put another way, either solar advances along some of the lines above, or significant climate change is on the cards. For the fossil fuel industry this must be the main threat.

New storage technologies

Even with a series of breakthroughs with direct solar, there remains a major handicap. As already noted, the trouble with most renewables – solar and wind in particular – is that they are intermittent. Wind turbines depend on wind speeds. Too much wind and the turbines have to be shut down. Too little and they stop turning. Solar depends on the sun shining and its intensity. Unlike wind, solar at least has the advantage that the sun shines when power demand is likely to be higher, and not at night when demand is lower. Solar intensity is fairly predictable and, in particular, as the light fades in the evening it is predictable when alternative power sources will be required.

Conventionally, the solution to intermittency is back-up provided by fossil fuels – large-scale coal and gas, and smaller-scale diesel and gas peaker technologies. For solar, there can be a reasonable ramp-up (and down) time since we know when it gets dark, so it is less reliant on the fast start-up diesels and gas peakers. The challenge in winter is obviously greater. For wind, back-up may be needed more immediately and randomly.

Back-up fossil fuels pose an obvious barrier to decarbonization. They are pretty indispensable unless electricity is stored. But if electricity could be stored, then the problem would go away. That is what makes the development of new storage technologies so critical, and potentially transformational for most renewables.

The few storage options available to date tend to be expensive. Pump-storage is one option: pumping water uphill when electricity supply is plentiful, and releasing it when needed. Dams hold back water and the reservoirs are in effect stored electricity. Energy can also be stored as heat. But none of these options represents the sort of really large-scale *and* decentralized *and* flexible storage that would be required with more and more intermittent electricity generation.[8]

The obvious answer is batteries. Without batteries most of the electronic devices we now rely on would not exist. There would be no iPhones and laptops. Current battery technology is limited in its capacity and by cost, but this may be changing. As and when it does, it is possible to imagine batteries everywhere – from a battery next to a wind turbine, in the house (possibly instead of a boiler) and in the car. Battery power is the route to the electrification of heating, transport and to the greater competitive penetration of intermittent renewables.

What are the chances of this happening soon?[9] There are two lines of development: enhancing and improving the now conventional lithium battery; and wholly new forms of batteries and storage using other vectors.

Conventional lithium-ion batteries have been around for a long time, but were only first seriously commercialized by Sony in the early 1990s. They are pervasive in electronics and therefore there is a lot of knowledge and research embedded in them. They are the obvious place to start looking for a step change in capacity. Yet lithium-ion technology has not so far proved amenable to the sorts of rapid improvements seen more generally with computers and chips. Moore's Law apparently does not apply (and neither does Swanson's).[10] Part of the reason is that lithium-ion batteries are very simple. Advances are largely about enhancing electrodes and electrolytes.

Quite striking claims have nevertheless been made about their potential. Two main lines of development are lithium-air batteries, using electrons partially from oxygen, and using lithium hydroxide rather than lithium peroxide. To these can be added the lithium-superoxide battery. All of these hold out the prospect for a steady technological evolution. It is obviously too early to tell which if any of these will deliver a step change. But serious money has already been invested, and household, car, smartphone and computer lithium batteries are being manufactured on an increasing scale. The Tesla Gigafactory in Nevada will give a further impetus to the conventional lithium-ion battery. Given the scale of the companies and industries committed to lithium, it has a significant head start. Big car companies plus big smartphone and big computer companies make a big difference.

The alternative is to try to find a completely different battery. The list of alternatives is bewilderingly large, and includes solid-state batteries, sulphur-based batteries and sodium-ion batteries. Then there are fuel cells. More typically targeted at cars, they might also be amenable to smaller consumer electronics, like smartphones.

There are already many possibilities and some look promising. It is a matter of when, not if, they become more effective and further breakthroughs are made. In addition to the impacts on cars and oil demand, widespread battery technology would not only change the economics of intermittent renewables, but also provide a way of balancing electricity systems at the grid level.[11] The result would be that the peaking and flexible fossil fuel technology might not be needed. Grid scale and local distributional system storage might mean that fossil fuels could eventually be almost completely eliminated from the electricity sector. When storage is combined with smart meters and grids (see below), the nature of networks and grids, as well as the role of flexible generation, will be transformed. Some even argue that high-voltage grids will not be needed at all.

In helping to address the problems of intermittent generation, storage at scale would further reduce the demand for gas and coal to generate electricity. This in turn would reinforce the tendency for fossil fuel prices to fall in the medium term, and provide another reason why our first predictable surprise would continue to play out over the medium and longer terms.

New transmission and distribution technologies

Intermittency can also be addressed through interconnecting electricity systems, both in specific circumstances and over longer distances. Whilst the wind might not blow in one area, it will somewhere else. Provided the interconnections are long enough to traverse different weather systems, and in particular beyond high-pressure–low-wind ones, the problem can be considerably ameliorated.

Long-distance interconnection allows renewable-generation technologies to exploit differences in more general weather patterns. Britain is a windy location – much better than central and southern Germany. Spain is a sunny country, much less gloomy than Germany. Wind in Britain and off its coast, and solar in Spain, make a lot more sense than lots of wind and solar in Germany. Like trade more generally, interconnection makes specialization much more efficient.

Interconnectivity works at the local level too. Distributed generation on a small scale within local networks relies on local small-scale back-up. The more robust distribution networks are, the better the economics of specific small-scale generation, from small-scale wind to solar panels and small flexible generation. Here technical progress is not just about the wires, but also the smartness of the network, at the smart network level. We return to some of these opportunities below.

As with generation technologies, there has been remarkably little progress in transmission cabling over the last century, either for local distribution or long distances. The once-trumpeted prospects of superconductivity have not materialized, and conventional wires continue to provide the backbone of the networks.

This is about to change in several ways. First, there are advances in conductivity and the materials used in transmission and distribution cabling. Second, there are advances in the use of direct current (DC) over long distances and then the ability to transform back to alternating current (AC). This is particularly relevant for offshore wind and interconnections between countries. Third, there are advances in the ways in which the electricity is sent through transmission systems. Finally, there are new and exciting possibilities which do away with the cables and use wireless technologies.

As with the developments in storage and generation, it is not possible to predict the path for breakthroughs, and this is not the place to try to review opportunities in any detail. The predictable surprise is that there will be breakthroughs; the less predictable surprise is which particular kind. The demand is there, and hence so is the scientific and business attention. Yet in many ways, this is the one area where they may not be necessary for a broader transformation to take place. The main obstacle to the development of transmission and distribution comes not from the limitations of the technologies themselves but the public acceptability of the transmission lines, and the costs of undergrounding these to avoid the visual intrusion, at least in developed countries. It is only recently, for example, that national electricity markets in Europe are being fully interconnected, and significant sub-sea cabling is being put in place. This has great potential. Iceland can bring geothermal power to Europe; Norway, and Scandinavia generally, can bring large-scale hydro to Northern Europe; gas can be brought from Russia and Norway to Western Europe; and solar can be sent north from Spain and even North Africa into France as French nuclear backs up Germany.

New broadband energy-consumption technologies and smart meters

Complementing the new generation technologies and the development of battery storage and improvements in cables, the demand side is about to change radically. To date, demand for electricity has been largely passive. System operators manage the supply side to balance the fluctuations in demand, and this is one of the reasons identified above for the vertically integrated structures that have dominated, and continue to dominate, the electricity sector.

There are two main reasons why demand is passive: metering has been crude, so there are gaps in the information; and there has been little or no interconnectivity between appliances and meters. This is why smart approaches are transformational: they allow for real-time monitoring of demand and real-time prices for consumers, and they enable smart appliances.

Some have argued that this will empower consumers to manage demand by looking at the prices and their consumption patterns. There may be an element of truth here for those who have the time to keep checking the meter. But the really radical changes come through active management systems, which do the job automatically for passive consumers. The white goods and appliances in the house can be linked to the broadband hub and use the smart meter data. This smart system can then moderate supply against price spikes, just as car computers now actively manage a host of aspects of car management.

The key to demand management becomes the broadband hub, of which the smart meter is a part. The hub provides the informational link to everything going on in the house and the outside world. The broadband hub can manage the services to optimize household energy use, and it can be augmented by not only smart appliances but household batteries, including the car battery as the take-up of electric vehicles unfolds. With household electricity generation too, the integrated decentralized network becomes a possibility. Householders can generate power; store surpluses in their batteries or sell to the network; take electricity from the system to charge batteries; optimize their use of appliances; and integrate their car battery and hence transport too.

None of this is going to happen quickly, and it will not be comprehensive. But incremental changes in this direction are already impacting on peaks, prices and behaviour. This has two radical impacts for the electricity industry: it enables the peaks in the system to be better managed; and it allows a host of technology companies to enter the electricity industry via the broadband hub.

Smart meters create an enormous amount of data. They enable the system to be optimized and the system operator to call on the demand side to help manage the balancing of the networks. As with batteries, this reduces the role of fossil fuels in providing back-up capacity when there are peaks in demand and when renewables are not delivering. If the wind is light, or it is very cloudy, the system operator can use the smart meters to turn down the marginal and non-time-critical demands. It makes sense to recharge batteries when generation is plentiful and to ease back when it is scarce, not just for the individual household generating electricity, but for the system as a whole.

Some customers might place a very high value on a really secure supply for all their uses. This is going to become more important as the economy digitalizes. Others might be much more price-elastic in their demands, and may have invested in the sorts of appliances that can take a more variable quality of supply. Some may be willing to interrupt demand altogether at high price points. IT companies, financial institutions and critical systems will put more of a premium on security of supply than households (possibly). Large energy-intensive users might have flexibility over production. Tariff and contract structures can tailor offers to customers. The key point from a systems perspective is flexibility, and the greatly enhanced ability to absorb shocks to demand and supply. Smart technologies greatly enhance system security.

In addition to enabling better systems management, the broadband hub with smart meters opens up new competitive opportunities, and brings a second radical change with it. The way consumption is organized in households is not just an electricity issue. Household behaviour has its complexities, something which the data from the smart meters (and from other sources such as web usage and Google searches) reveals. There is now a lot of big data to work on, and the speed of data generation is extraordinary: 90% of the world's data was created in the last two years, and the path is exponential.[12]

The companies that may be best placed to exploit these opportunities are probably not going to be conventional electricity supply companies. They are more likely to be new players with world-class IT capabilities, including the likes of Amazon, Google and a host of app developers. As we shall see in Part Three, this has major implications for the incumbent electricity companies, and will drive a further stake into the heart of the conventional vertically integrated industry structures. The fact that the marginal costs of the energy provided may be zero adds to the threat to the incumbents' conventional model. That too is a predictable surprise.

Electric cars and hybrids

Next up in this fast-changing technological landscape is electric cars, with major implications not just for the electricity sector but oil too, since transport has the biggest share of oil demand. These are already making their presence felt, and the electrification of cars is only one of the major changes happening in the transport sector. Cars are becoming not just electric but smart too. They are increasingly semi-automatic, with functions from handbrakes to parking being taken over by the car's computers. Eventually they may be driverless altogether, effectively becoming personalized transport pods. Trucks are already there: in major mining and other off-road industrial complexes their presence is increasingly normal, and they are being trialled in convoys on major roads.

Costs are still relatively high and there are no sophisticated charging infrastructures, so progress has been slow. Without proper pollution pricing, especially from diesel cars, the internal combustion engine maintains an unfair advantage. Potential customers worry about the limits of the battery range, notwithstanding that most journeys are in towns and cities and well below the range limits, at least in urban conurbations and more densely populated countries.

This is about to change. The big car companies are moving in with large-scale investments (while at the same time hedging their bets by exploiting the sports utility vehicle (SUV) markets as oil prices have fallen). Batteries are becoming more advanced, as described above. Gradually, the take-up, encouraged by urban planners worried about congestion and air quality, is reaching the threshold where charging systems are rolled out. Hybrids, fuel cells and hydrogen are all part of the transitionary process. There will be several tipping points, both in terms of fashion and social acceptability, and in terms of volume, cost and infrastructure. It is not yet clear when, but it is no longer an 'if'. Lower oil prices do not change these dynamics, even if they slow them down, because transport is increasingly about much more than the pure cost of the fuel.

Imagine what a fully electric car transport system would look like. The demand for oil for transport would fall away, and with it part of the demand growth the peak oil theorists rely on. The electrification of transport would seriously dent oil demand growth and, given that oil supply is expanding with shale technologies and other developments described in Chapter 1, prices would go down rather than up. Lower oil prices slow the transition, but they are unlikely to see it off, especially if governments step in with carbon prices, the regulation of air quality and other inducements. The twentieth-century story of oil for transport would morph into a twenty-first-century story of electricity for transport.

In the short term, the substitution may be between oil and gas via electricity generation. Gas-fired power stations can contribute towards generating the electricity for charging the cars. So the golden age of gas would have a further boost from electric cars, and the geopolitics of energy for transport would shift from oil to gas in the first instance. Gas can also directly power trucks, as has happened in the US. Further ahead, it may be replaced by next-generation solar and other renewables, as described above. The path goes from oil to gas to next-generation renewables.

Electric cars have other disruptive system characteristics. Cars currently store lots of fuel. Think of the aggregation of all those half-full fuel tanks. The importance of this stock can be seen when there are threats of interruptions, from strikes by tanker drivers, for example. Panic buying rapidly exhausts storage in the rest of the system, as filling stations and refineries run out very quickly. Now imagine that this storage is of electricity instead. It would have an important impact on the electricity systems if it could be managed, and especially if linked into the smart technologies and broadband hubs described above.

Cars could be recharged at night when demand is low; they could be recharged at high solar- and wind-generation times. Smart technology in the car, which can 'remember' the driver's usage habits, could allow this electricity store to be managed for the benefit of the system as a whole. The car-charging could also be optimized against the household consumption patterns, and even household batteries.

These electricity-management opportunities help to change the nature of peaks and the need for peaking and standby capacity. Electric cars represent not only a new demand for electricity, but also a fundamental shift in system management. As with the broadband hub and smart meters, it is not at all obvious that the existing vertically integrated incumbents would be especially good at competing in the electric car markets. New entrants might instead come from major car companies – indeed they already do. The interface between car companies and Google and Amazon on the one hand, and renewables generators on the other, is yet to be determined.

New materials

It is not just the generation, storage, cabling and supply markets that are being radically changed by new technologies. The very basis of manufacturing, and the materials used, are also undergoing a transformation. These technological changes are likely to have profound impacts on our energy future, and are as yet very poorly understood.

Materials for our industrial world come down to things like steel, cement, plastics and other petrochemicals. As with the energy sector more generally, there has been remarkably little change in this make-up. For almost a century, technological developments have focused on making incremental improvements to these rather than coming up with new materials – for example, by adding more elements, from the early use of tungsten to harden steel, through to a range of rare earths used in modern wind turbines, mobile phones and a host of critical products.[13]

Wholly new materials like graphene could change the game from incremental advances to step changes. Graphene combines a number of very attractive properties. It is a single layer of carbon atoms; it is very thin, very strong and very flexible. It is 200 times stronger than steel, and a million times thinner than a human hair. In addition, it has very good electrical properties: it is electrically and thermally conductive, and it is optically transparent, making it a candidate for developing new solar cell technologies. It has battery applications, potentially making a step improvement in charging lithium-ion batteries. More indirectly relevant to energy demand, it has lots of applications to composites, membranes and semiconductors. It would be surprising if other new materials were not developed during the remainder of this century.

These new materials have two potential applications relevant to the future of energy: to manufacturing generally, and to electricity specifically. On the general front graphene might be a substitute for other materials. It could substitute for plastics (and hence help in displacing oil- and gas-based petrochemicals) and other materials in construction and the production of cars and aeroplanes, for example. Consider how plastic has transformed our world. It is now ubiquitous – so much so that it has polluted the entire planet, from the plastic bags that litter the countryside to the great waste gyration in the Pacific and on the shorelines of Antarctica and the Arctic. It is, along with cars, one of the greatest practical manifestations of the age of oil. Now imagine a replacement material. A shift in materials away from plastic would seriously damage the demand for fossil fuels, and especially oil. Yet that is exactly what decarbonization requires.

Will new materials emerge? Almost certainly, there will be very predictable surprises. Will graphene be 'the big one'? It is far too soon to tell. It could be yet another over-hyped short-lived wonder, or it could be transformational. On the plus side, graphene has already resulted in an explosion of patents (especially in China and the US). It has attracted lots of public funding, and the sheer multiplicity of applications that have been touted suggests that it has the virtue of being a general enabling technology. On the downside, there are few common standards; new start-up companies have yet to bring significant products to market (though there is now a graphene-based light bulb and a graphene condom has been developed[14]); and it will probably take years for graphene to be mainstreamed into industry, if ever. The gradual demise of oil does not require graphene's advocates to be right about its particular potential, but it will require new materials to displace petrochemicals.

New industrial production technologies: 3D printing, robots and AI

The coming of new materials is just one of the ways in which manufacturing is changing. A second impact comes from the application of information technologies. IT enables 3D printing by digitalizing objects. The digital image can be manufactured by printing out layers of materials, typically in the form of gels and powders. What is radical about 3D printing is the ability to customize each item from the desired digital image. It is the opposite of the mass production techniques of the twentieth century. Instead of standardization to reap economies of scale in vast factories on a global basis, a 3D printer can be localized and specific. It can be used in the kitchen and on a building site. Even houses can be 3D-printed.

As with graphene as a potential general-purpose material, it is too early to tell for sure how this technology may change economies, though in the case of 3D printing we can be pretty certain that it will. Indeed it already is. It will impact on the amount and location of electricity demand, and it might favour decentralized generation. Manufacturing can be located closer to final demand. This could have radical implications for countries like China, built on mass production at a long distance from the ultimate customers (as we shall see in Part Two).

Robots and robotic technologies are bringing about a second transformation of manufacturing – what Erik Brynjolfsson and Andrew McAfee call the 'second machine age'.[15] These are again the product of digitalization and IT. Robots not only make manufacturing much more capital-intensive, and displace labour, but they also improve efficiency. They are not time- (or sleep-) dependent, and their productive activities can be scheduled against the fluctuations in electricity supply and demand. They could have a significant impact on peaks, and flatten out the profile of demand, supply, and hence prices. The need for balancing peaking capacity may be offset by more flexible manufacturing based on robotics.

The conventional idea of a dumb robot responding to human commands is giving way to much more radical robotic technologies, integrated with advances in AI. With these advances comes massive data, and the possibility of new computing technologies. Quantum computing technologies are being pursued by all the main global companies, including Google with D-Wave and Microsoft with Station Q. These technologies replace the conventional binary zeros and ones.[16] Robots that can learn, and that can operate with a degree of autonomy, are beginning to take over a whole host of functions. The driverless car is one form of robotic technology, computing massive amounts of data. Drones offer new delivery and surveillance systems. Nanobots offer major medical breakthroughs.

Among the many applications, almost every stage of the energy chain is open to the deployment of these new technologies, from sensors and exploratory devices for fracking and conventional well management to nuclear safety inspections and repairs, through to line fault management and maintenance.[17]

3D printing and robots are just examples of what the IT revolution has facilitated. IT is one of the most pervasive general enabling technologies, on a par with the invention of printing and electricity. These technologies change everything – eventually, every production and consumption process in the economy. They take time to play out. This one is faster than any in the past. Whereas printing and electricity each took a century, IT has already affected almost everything. In just thirty years we have gone from the typewriter and fax machines to laptops, smartphones and Internet banking.

There have been direct impacts for the energy sector, notably on trading and managing systems. Liberalized traded wholesale markets and retail switching would not have been possible in the 1970s and early 1980s. But it is on the system side that the impacts are likely to be the biggest. All networks are systems, and systems lend themselves to IT – from trains and cars to water, energy and communications. They are all energy-intensive, and all have major locational impacts. The development of smart ways of utilizing assets, of sharing consumption (for example, the online transportation network company Uber) and splitting ownership from use, will change the very nature of the economy, and hence future energy demand patterns.

New technologies and the demand for fossil fuels

Almost all of the new technologies point towards electricity and its growing dominance. In Part Three we shall examine in greater detail the winners in our energy future. But what will be the impact of these new technologies on the potential losers – the oil, gas and coal industries?

For oil, its key markets are negatively impacted. Electric vehicles are beginning to attack oil at the margin. As long as the electric share grows more slowly than the growth of car ownership, transport will continue to push up oil demand. But the growth path for electric cars is unlikely to be linear. Urban pollution regulation will speed the process. The tipping point is likely to happen in the next decade or so, at which point more cars do not translate into more oil demand. Eventually electric cars will start to reduce oil demand.

The petrochemicals side of the oil market is a derived demand – for plastics, fertilizers, cosmetics, detergents, synthetic cloths and asphalt. Plastics play a huge role in packaging, followed by buildings and cars and trucks. These products can be manufactured from any fossil fuel or biofuel, but until recently, oil in the form of naphtha and condensates has dominated ethane, propane and butane. In 2014, the ratio of the former to the latter was nearly 70:20. In the US, this ratio has shifted significantly since shale gas came on the scene.[18] A key conclusion from this chapter is that oil is vulnerable: not only is gas a substitute as a fuel input, but new materials like graphene might also start to cut into the demand for oil via their impacts on the demand for plastics.

Taken together, new technologies may mean that the demand for oil may soon stop rising and then begin to fall back – before the climate change considerations are taken into account. The radical implication is that in the medium-to-longer run the price of oil will probably fall because of technical change, as well as because of the carbon constraints, and the increases in supply.

The demand for gas is much harder to project. Gas has many advantages in the new-technology world, at least in the short-to-medium term. It is a much better environmental option than coal in electricity generation, and for this reason its demand is likely to go on rising. Cheaper gas in the US also means that it has a bigger role in the petrochemicals industry. But the role may not last into the long term: solar generation might outcompete gas even before the need to phase it out for climate change reasons, and plastics and other petrochemicals products may be challenged by substitutes. The golden age of gas – effectively since 1990 – may be brief, and certainly much shorter than oil's century-long hegemony.

The demand for thermal coal is also tied to electricity and at first sight more electricity might result in more coal demand. But phasing out coal is now an explicit climate objective and an energy policy objective, notably in

Britain. This is likely to take effect – and lead to a switch to gas – before the new technologies can gain a competitive advantage. It is a matter of policy, not markets and prices, and regulation is the main policy instrument.

Coal for industrial purposes is a derived demand – derived from the use of coking coal in making steel (globally, 70% of steel produced today is made using coal) and also for aluminium, cement, paper, chemicals and pharmaceuticals. Then there are the by-products from making chemicals and ammonia for fertilizers. New materials cut into this demand.

In addition to these specific fuel effects, the new technologies will have considerable demand-side implications, and affect the level and shape of energy demand. Energy efficiency will improve as new transmission cabling and smart technologies come on-stream. This will not, however, necessarily reduce total energy demand: indeed an increase in energy efficiency is equivalent to a fall in price and falling prices tend to *increase* demand. This is what has been going on for the last 200 years. As long as it is low-carbon, increasing demand is a 'good thing'. The electrification of transport will increase electricity demand, as will the more general shift in manufacturing towards electricity. The volatility of demand will be flattened and storage will further contribute to reducing the demand for peaking services. Driverless cars may add further to demand: older people will be able to be driven by computers when they are no longer capable of driving themselves. Robots will provide a host of new services, many of which we probably cannot even imagine yet, but all requiring electricity.

Taken together, these technological improvements are likely to increase the demand for energy in general, and electricity in particular, over the coming decades. The conventional assumption that the demand for energy and economic growth has decoupled may not survive these changes. More demand does not mean that prices will rise, however: it depends on the characteristics of this overall demand, and on the specific demand for the particular energy sources, and it also depends on supply. Recall from Chapter 1 that a century of rising demand for oil up to the 1970s led to a gradual *fall* in the price. If next-generation solar turns out to be very cheap – if and when more of the light spectrum is opened up and new applications like solar film are developed – then the greater demand may be accompanied by falling electricity prices. Saving energy may not be necessary after all.

It can't be stopped

With so much going on, why do so many environmentalists fear technology and deny its central role in decarbonization? There are at least four major reasons. First, there is a general technophobia, which comes from long exposure to its many unintended consequences. Past technological developments have brought pollution, and most of the twentieth-century pollution has been the product of technological progress. Think of the destruction of biodiversity, the plethora of wastes, and the changing of the atmosphere ushering in perhaps dangerous climate change. Second, there is a military fear that new technologies have given rise to ever-greater powers for destruction on a planetary scale. Nuclear weapons are just one example, and the one that many environmentalists fear the most. Add drones, military robots and the possible applications of AI, and a whole new type of conflict becomes possible, both by world powers and small terrorist groups. Third, there is a resource fear that, however sophisticated technologies might become, they still depend on (scarce) natural capital. The low-carbon technologies will still rely on other (sometimes rare) metals. Finally, there is the deep ideological argument that ever-greater technological capabilities will underpin rising populations and higher economic growth rates, destroying the fabric of the earth's ecosystems.

At the very least these factors could slow down the take-up of new technologies and raise the costs. Nuclear energy and genetically modified crops are examples of where this has happened already. The NGOs and environmental campaigns have seriously hampered both.

The fossil fuel industry and the major producer countries might also use their lobbying power to slow down the technological transformations described above. Technological progress is not only low-carbon. It caused the shale revolution and it has been a causal factor in bringing the commodity super-cycle to a shuddering halt. Lower prices make it more competitive. Some countries might play to this advantage and undercut those countries transitioning away from the fossil fuels.

Yet this time it is already different. The oil companies no longer have it all their own way. The new companies of the information and digital era counter them. Google and Apple lobby against them. New renewables interests have sprung up, and their grip on the media has displaced that of the fossil fuel industries. Even financial institutions have begun to migrate away from past support. Low prices make life tougher, reducing dividends and diminishing the scope for capital gains. Then there is the 'moral' divestment movement and associated political and social pressures to shift away from investing in fossil fuels.

But could the combination of environmental technophobia and Big Oil derail a transformation of the energy sector? Fear of driverless cars might delay large-scale deployment. Luddite approaches to robotics might stymie their use. Yet these are all quite limited constraints. For all the fears, the main technologies – smart grids and meters, next-generation renewables, new materials, robotics, 3D printing and AI – look unstoppable. They will almost certainly transform our future economies and the demand and supply of energy. This is the third and most important predictable surprise. For some, like the US, Europe and Japan, this looks like good news. For the oil producers and in the Middle East and Russia it is bad news. For China the energy situation may improve, but in a world of robotics, 3D printing and AI its cheaper labour no longer looks like such a world-beater, and the jury is out.

The remainder of this book considers the consequences of all three of our predictable surprises. It is a terrible story for the producer countries, and a better story for the developed consumer countries. It is very bad news for oil and coal companies, but there are great opportunities for the new electricity-based companies. Best of all it is very good news for the fight against climate change and the many other forms of pollution derived from fossil fuels. In a context dominated by environmental doom and gloom, there are very real grounds for optimism on the energy front.

PART TWO

The Geopolitical Consequences

The twentieth century saw the gradual economic and then political rise of the oil producers. First Russia and the US enjoyed early leads, and both industrialized, utilizing their own abundant resources to become super-powers. Neither could have made it so far without oil (and coal and gas). Then came OPEC, and the two bursts of dominance in the 1970s and in the 2000s for Saudi Arabia, Iran and Iraq, and boom years too for the second-division players like Algeria and Libya. This did not prevent the Arab Spring breaking out, and already the Middle East is ablaze with conflicts in Northern Iraq, Syria and Yemen, and with heightened tensions between Iran and Saudi Arabia played out in proxy wars across the region. The US, Russia and Turkey have been drawn in too. To make matters worse, the commodity super-cycle has collapsed.

The three predictable surprises described in Part One will play out for the OPEC countries over the coming decades in this historical context. These losers of the endgame for the fossil fuels will be somewhat offset by the winners on the consumer side. This part of the book is all about the impacts in store for the biggest countries on the producer and consumer sides of the energy markets. These are, on the producer side, the US, Russia and the major Middle Eastern players (Saudi Arabia, Iran and Iraq), and on the consumer side the US (again), China and Europe.

Others have significant parts to play, but not on the scale of the countries considered here. The US plus Europe plus China make up around 70–75% of world GDP, and the US, Russia and Saudi Arabia each produce over

10 mbd. They may be joined by Iraq and Iran eventually, and then these five would meet over 50% of total oil demand. India and parts of Africa, Southeast Asia and even Latin America are all potential energy consumer giants. Canada, Venezuela, Mexico, Argentina, Nigeria and Angola are all actual or potential producer players. They will each have a role in the endgame, but this is likely to be dominated by what happens in the countries and regions discussed here.

CHAPTER 4

The US
THE LUCKY COUNTRY

Like Saudi Arabia and the other Middle Eastern oil and gas producers, the US has abundant natural resources. Unlike them, it has not been cursed as a result – indeed, quite the reverse. Why has it been so lucky? And why, too, is it emerging at the end of the commodity super-cycle as the great winner – compared not only with the other oil-cursed countries (including Russia), but also China? Why will the gradual decline of the fossil fuel industries leave it in a very competitive position?

The US was in at the start of the oil industry; it has been the biggest player for most of its history; it has been the largest carbon emitter and, until recently, the world's largest economy. Its past is built on natural resources, and oil transformed the US economy. The early settlers had a cornucopia of land, water, timber, animals and minerals in front of them – the Frontier – and the recent development of shale oil and gas is just the latest chapter of its extraordinary history. Nature's endowment has bequeathed enormous natural capital to the lucky Americans. With a relatively small fraction of the world's population, they have stripped the land of its forests, cultivated the Great Plains, decimated the native fauna and expropriated the native peoples. They have extracted lots of fossil fuels and begun to change the global climate.

Americans have built the richest and most powerful economy in the world, and harnessed an enduring lead in technology. Its great corporations are globally dominant – from the Exxons and Chevrons of the oil industries to the Microsofts, Googles and Apples of the Internet age. Most recently, the US has developed fracking technology on a scale and speed that no other country can remotely replicate. Whether the US can now lead in the new low-carbon technologies is a matter on which the global climate may hang. Because our climate will depend on these new technologies, as we saw in Part One, it has a better chance than most.

The lucky country has now not only become the technological hub of the world, and developed shale, but it is also benefiting from gradually weaning its economy off dependency on energy imports. Its natural capital is so great that it has made significant strides towards energy independence. Not only will this transform the US trade position (until recently, half of the deficit was due to energy imports), but may also remove the US's dependence on Saudi Arabia, the rest of the Middle East and OPEC. The significance of this, for the US and for geopolitics, has yet to be fully appreciated, though it is already sending shudders through Saudi Arabia and the Gulf states.

All this makes the US the most likely winner in this new energy context. Lower oil prices help its economy; its domestic production cuts its trade deficit; and the new technologies play to the US's strengths. Even better, the digital technologies undermine its Achilles heel in terms of global competitiveness – the challenge from cheap labour in developing countries. Even on the carbon front, its domestic shale gas makes it easier to cut into its emissions from coal, and therefore to adopt carbon targets that it can achieve at little extra cost.

The US's oil century

The US got into oil right at the beginning, in the second half of the nine-teenth century. It was yet another new Frontier. Consistent with the broader US history of pushing the land frontier ever westwards, the oil frontier was all about getting the black gold out as fast as possible, before anyone else. US property rights and free enterprise encouraged so-called wildcatters to drill wherever they could, and to reach as far horizontally as they could. This set of incentives would again underpin the shale oil and gas revolutions almost a century later.

The US addiction to oil gradually unfolded alongside the love of the car. The two went hand in hand, and still do. Americans have a deep aversion to gasoline price shocks. In a country with wide-open spaces and limited rail and other public transport, America celebrates the car in a way few others do. Figure 4.1 shows the intimate relationship between the oil industry and car ownership.

US production soon proved insufficient to meet growing demand, and after the First World War, encouraged by government, US companies moved towards augmenting domestic supplies with imports. The scale was initially small, however, as Figure 4.2 indicates. It was not until after the Second World War that import dependency really took off, and not until the peak of US production in around 1970 that imports became critical.

As the reliance on imports grew, so too did the spread of the US oil companies, and therefore US foreign policy on the back of those oil imports. The first US imports came from its backyard: Mexico, Canada and Venezuela. US foreign policy had already asserted dominance over the Caribbean and the Gulf of Mexico with the Monroe Doctrine.[1] The really big breakthrough came with Saudi Arabia, though it was a gradual process, kicked off by the concession of Standard Oil of California (SOCAL) in 1933. After the Second World War, the growth of imports and US oil dependency was a story overwhelmingly of demand growth, as the golden age of

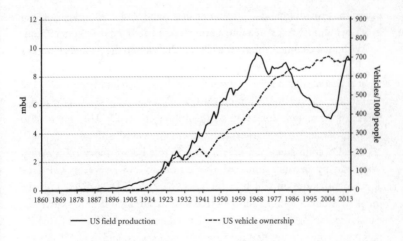

Figure 4.1 US crude oil production vs vehicles per 1000 people

Sources: EIA, 'U.S. Field Production of Crude Oil', 31 August 2017, http://www.eia.gov/dnav/pet/hist/LeafHandler.ashx?n=PET&s=MCRFPUS2&f=A; Davis, S. C., Diegel, S. W. and Boundy, R. G., *Transportation Energy Data Book*, Edition 35, Oak Ridge National Laboratory, October 2016, cta.ornl.gov/data

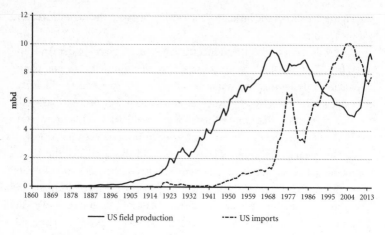

Figure 4.2 US crude oil production vs imports, 1860–2016

Sources: EIA, 'U.S. Crude Oil Imports', 31 August 2017, http://www.eia.gov/dnav/pet/pet_move_impcus_a2_nus_epc0_im0_mbblpd_a.htm; EIA, 'Crude Oil Production', 31 August 2017, http://www.eia.gov/dnav/pet/pet_crd_crpdn_adc_mbblpd_a.htm

economic growth unfolded. A quarter of a century of economic growth at 2–3% per annum translated into a growth of oil demand of 3% per annum.

Without a colonial empire, the US was drawn into the Arabian Gulf as a result of the First World War. It had no previous form in the Middle East – an area that at the time was dominated by the French and British. Oil, and US oil dependency on the Middle East, would eventually see it as *the* major power in the region, along with a large military presence. The long road to the Iraq war and to the Syrian civil war was one paved with oil.

The US companies had only a limited role before the Second World War, playing second fiddle to the European powers. The Red Line Agreement (more on which in the next chapter) in 1928 divided up the former Ottoman Empire between them for exploration. Saudi Arabia was considered unlikely to hold any significant deposits and it was left to the US companies outside the Red Line Agreement (Texaco and SOCAL, later to become Chevron) to go after Saudi Arabian exploration.[2]

The post-Second World War US economic miracle was energy-intensive, and this presented little problem while oil prices remained low and stable. From the US perspective, the Middle East remained a political and military

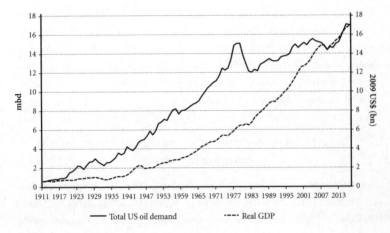

Figure 4.3 US crude oil demand vs GDP, 1910–2016

Sources: EIA, 'U.S. Crude Oil Imports', 31 August 2017, http://www.eia.gov/dnav/pet/pet_move_impcus_a2_nus_epc0_im0_mbblpd_a.htm; EIA, 'Crude Oil Production', 31 August 2017, http://www.eia.gov/dnav/pet/pet_crd_crpdn_adc_mbblpd_a.htm; Johnston, L. and Williamson, S. H., 'What Was the U.S. GDP Then?', MeasuringWorth, http://www.measuringworth.org/usgdp/

backwater, with little to threaten what appeared to be secure and plentiful supplies. The US continued to be involved in wars elsewhere in Southeast Asia – first in Korea and then in Vietnam. But in the Middle East, not even the Suez Crisis in 1956 threatened US suppliers. The US, as well as the Europeans, continued to rely on Saudi Arabia and Iran, and when Iran's internal politics threatened the stability of oil markets, the US (with British help) overthrew the short-lived Mosaddegh revolution with a coup in 1953, and (re)installed the US-friendly Shah. If the US had an interest, it was more about the role of the Russians in Iran, Egypt and later Syria, rather than a fear of oil embargoes and price hikes. All this was yet to come.

This easy complacency sowed the seeds of the disasters to follow in the 1970s. Israel remained the US's Middle Eastern focus, and the Arab determination to protect the Palestinians meant that no equilibrium could be achieved. What made this regional conflict between Arabs and Israelis toxic for the oil market was what followed from the joint military attack on Israel by Egypt and Syria in October 1973. In response to the US decision to resupply the Israeli military, OPEC imposed an oil embargo, which it then extended to other countries that supported the Israelis. Finally, for good measure, it cut production too. For the US, this was a further shock in the context of a greatly deteriorating economic situation. The long post-war boom was already over, ending in the abandonment of the Gold Standard and the Bretton Woods Agreement, and kick-starting a decade of stagflation. It was three disasters in a row.

The OPEC oil shocks of the 1970s and the US's first attempts at energy independence

The 1970s OPEC oil shocks were devastating for the US. Oil prices quadrupled from $2.90 before the embargo to over $11 immediately afterwards. OPEC not only ramped up prices, it also allowed the political dimensions to spill out into a new and much more fundamentalist backlash against the West. By the end of the decade the Iranian Revolution split the Sunni and Shia worlds, led to the Iran–Iraq war, and changed the game for ever. Whilst there had been Middle Eastern disputes for the US to deal with before, these were largely about the terms of the oil contracts and the distribution of the economic rents (and of course Israel). After 1979, they gained a hard ideological edge. The US would find itself involved in two Gulf Wars, and the region would descend into a chaotic unwinding of most of the remaining assumptions of the post-Ottoman world. On 11 September 2001, the terrorist attacks on the US, mostly carried out by Saudi nationals, created some of the most iconic photographic images of the modern age – particularly the destruction of the Twin Towers – and in the process further undermined US confidence in the Middle East.

As the balance of power shifted with the growing US dependency on oil imports, nationalization was the obvious route for the producers. The NOCs would go from minor players to dominating world oil production. Where once Americans had developed and operated oil fields and wells, now as 'infidels' they were even at times kept off Saudi soil. The great days of Aramco and the US companies were over.

The OPEC shocks of the early 1970s had global consequences. As noted, they caused a global economic recession combined with inflation – stagflation – in a context in which the Bretton Woods currency system had already ended with the free float of the US dollar and its subsequent devaluation. US and European economies struggled to cope with the fallout, made still worse by the doubling of oil prices in 1979. The scale of pessimism about the US's economic prospects is hard to comprehend now, but the combination of a widespread belief that the world was running out of key minerals, the scale of the price shocks, and the recognition of dependency led to a profound sense that the 'oil addiction' had to stop.

In the early 1970s, President Richard Nixon faced a daunting foreign policy challenge: how to carry on supporting Israel, keep the US's Arab oil suppliers on side, and keep Russia out. Although Secretary of State Henry Kissinger managed to fudge these trade-offs sufficiently to get the embargo lifted by showing that the US was at least trying to bring peace, he did not solve the problem, and neither did any of his successors. Although other complicating dimensions have been added (in particular, the Iran–Saudi Arabia split), all these challenges remain today.

Nixon did the other obvious thing: he launched a new energy strategy to reduce import dependency, and in the process kicked off the desire for more energy self-sufficiency – Project Independence. As he expressed the ambition:

> Let me conclude by restating our overall objective. It can be summed up in one word that best characterizes this Nation and its essential nature. That word is 'independence'. From its beginning 200 years ago, throughout its history, America has made great sacrifices of blood and also of treasure to achieve and maintain its independence. In the last third of this century, our independence will depend on maintaining and achieving self-sufficiency in energy.
>
> What I have called Project Independence 1980 is a series of plans and goals set to insure that by the end of this decade, Americans will not have to rely on any source of energy beyond our own.[3]

The energy independence goal would be taken up by his successors, and in particular by Carter, and would resurface with George W. Bush's Energy Plan in the 2000s. Since they had no way of knowing that shale oil and gas would one day 'solve' the problem, in practice independence meant getting away from oil, and finding alternative supplies beyond the grip of OPEC.

The goal of energy independence is easy to state, but in practice there were few policy levers. US oil production peaked in the 1970s, and this 'peak oil' was taken as a permanent weakness of the American economy. As we saw in Chapter 1, Hubbert's prediction had turned out to be right. Not much could be achieved quickly on the supply side, other than the creation of a Strategic Petroleum Fund and encouraging the creation of the IEA.[4] There was little spare capacity in the US oil fields (notably in East Texas) to boost home production, so inevitably the demand side was the initial focus – energy efficiency and measures to reduce demand. A speed limit of 55 miles per hour was introduced in 1974, to be followed (after Nixon resigned over Watergate) by the introduction in 1975 of the Corporate Average Fuel Economy (CAFE) standards under Gerald Ford.

There began a search for alternative energy sources. Nuclear offered one answer for electricity generation, combined with coal. Carter shared Nixon's (and Ford's) mindset on energy and, as we saw in Chapter 1, he was certain about a future of ever-higher oil prices. His energy policy is perhaps best remembered for his passion for solar, epitomized by his installation of solar panels on the roof of the White House. In his 'solar message' to Congress, Carter stated:

> On Sun Day, May 3, 1978 we began a national mobilization in our country toward the time when our major sources of energy will be derived from the sun. On that day, I committed our Nation and our government to developing an aggressive policy to harness solar and renewable sources of energy. I ordered a major government-wide review to determine how best to marshal the tools of the government to hasten the day when solar and renewable sources of energy become our primary energy resources. As a result of that study, we are now able to set an ambitious goal for the use of solar energy and to make a long term commitment to a society based largely on renewable sources of energy.[5]

But in fact the energy policy he had already set out in 1977 majored on the greatest possible utilization of coal in utilities,[6] reducing energy demand through energy-conservation measures, and banning gas from being used in power stations (the latter remained in place until 1990).[7]

The humiliation of the failed attempt to rescue the US embassy hostages in Iran sealed the sense of vulnerability. Following the seizure of the US embassy staff in Tehran in November 1979, Carter embargoed Iranian oil and started the sanctions game, which would last right through to Trump. He launched a rescue mission, Desert One, in April 1980. It went spectacularly wrong.

Even the nuclear option proved difficult. The Three Mile Island nuclear accident in March 1979, in which the reactor partially melted down in Dauphin County, Pennsylvania – an urban area – was an enormous setback for the industry, just as oil prices were transforming its economic prospects. The US nuclear industry never recovered, leaving coal in the driving seat. The actual human consequences were very limited but, as in all things nuclear, it is the perceptions that matter: the very small probability of a massive impact. Faith in the technology was seriously dented.

Yet against the background of all this pessimism, energy markets in the US were working. As Figure 4.4 illustrates, after a continuous climb since the Second World War, demand growth fell back in the 1970s (as it would again in the 2000s, albeit then not only in response to price but also because of outsourcing to China). High prices dented demand and encouraged energy efficiency, and they also motivated the search for alternative sources of supply. What mattered was less the policy objectives of Nixon and Carter, and more the key ingredient – price. Markets, not policy, drove outcomes in most countries, but especially in the US.

The 1970s may have yielded more money for OPEC, but the decade also spawned a new ideological radicalism, funded by oil and incubated in

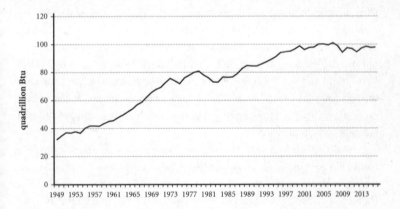

Figure 4.4　US primary energy consumption, 1949–2016

Note: Btu, British thermal unit.

Source: EIA, 'Monthly Energy Review', August 2017, https://www.eia.gov/totalenergy/data/monthly/pdf/mer.pdf

Afghanistan, which then spilled over into the Gulf states. For a thousand years, Sunnis and Shias had cohabitated in the Middle East, largely in peace. But then Sunni-led Iraq challenged revolutionary Shiite Iran with an invasion in 1980, in a bloody conflict that ended in stalemate in 1988. The great religious schism dating back to the succession following the death of the Prophet firmly divided the Muslim world.

Iraq was further undermined by the fall in oil prices in the mid-1990s. Saddam Hussein used the excuse of blaming Kuwait for over-supplying the market, and thus contributing to these lower prices, to justify invading it in order to rectify what he saw as the artificial boundaries drawn up after the Ottoman Empire collapsed, and in the process illustrating the extent to which the carve-up of the old empire had been crude and failed to reflect the ethnic and religious differences on the ground. The post-Ottoman legacy and the religious schism would later dramatically reappear as Syria imploded and ISIS took over parts of Northern Iraq.

The US goes to war, partly for oil

The US's response to the invasion of Kuwait reflected its own priorities. For all the propaganda about the evils of Saddam Hussein's dictatorship (and there were evils), the US did not go to war primarily for the benefit of the Iraqi people. Without oil there would have been no full-scale military campaign. Saddam's threat to Saudi Arabia was real, and the US had no desire to have a re-run of the oil crises of the 1970s. The fact that Desert Storm, the name given to the counter-attack, was backed by a UN Resolution, and the US assembled a coalition of thirty countries, does not distract from the central issue – oil. A long air campaign was followed by a ground offensive that took just 100 hours to finish the job, and put Saddam Hussein back in his Iraqi box.

The First Gulf War turned out more like a duck shoot than a modern conflict. The two sides – Iraq and the US-led coalition – were unevenly matched, resulting in a rout of the Iraqi army, but one that stopped at the Iraqi border, leaving Saddam Hussein in power. The Second Gulf War more than a decade later was a similarly uneven match, leading to rapid US success. But the lack of a credible follow-up plan, and the failure to understand the Sunni–Shia divide, meant that it effectively never ended, spilling over first into civil war between competing factions inside Iraq and then morphing into the ISIS takeover of the north, and becoming part of the Syrian civil war. Iraq had been at war with Iran from 1980 to 1988, and with Kuwait and the US in 1990–91, and it was then invaded in 2003 by another US-led coalition. With the US delivering southern Iraq and control over the national Iraqi government to Iran (which Iran had failed to win during the eight years of carnage of the Iran–Iraq War), Iraq has remained in a state of civil war ever since. It could easily become a twenty- or even thirty-year war. Indeed, it is plausible to argue that the Middle East has been embroiled in one long post-Ottoman Empire conflict.

For the US the lessons have been chastening. It no longer has the power to impose its will with ease on the Middle East, and has gradually sought to reduce its role. Having bombed Gaddafi's Libya in 1986, the US did not join in the Anglo-French air strikes to facilitate the removal of Gaddafi in 2011; the US was wrong-footed in the Egyptian Revolution; it refused to engage in Syria despite its formally stated 'red lines' on chemical weapons being breached by President Bashar al-Assad's regime; and it has been reluctant to engage ground troops in Northern Iraq to tackle ISIS. Then it was outsmarted by the Russian military interventions in support of Assad in Syria.

Win one: the great escape and the coming of shale

A core reason for this reluctance to intervene in Middle Eastern affairs is that the US is escaping the noose of dependency. Our first predictable surprise, and the role played by shale, is releasing the US from more than seventy years of increasing dependency. This is its first 'win'. As discussed in Chapter 1, the development of shale deposits in the US was a combination of price, a number of technologies and particular property rights. Shale developed extremely rapidly, and turned the US's perceived weakness in energy into a strength. Both Gulf Wars were conducted in the context of a fear of the dependency. Now the constraint no longer binds.

New facts take a long time to undermine old certainties. Yet the facts matter, and they are increasingly hard even for those most steeped in the Nixon–Carter–Bush energy outlook. Rough energy independence, Nixon's goal, is within reach. The US's primary vital national and strategic interest is no longer dependent on a small number of authoritarian regimes with bloody pasts and presents. It has the option to draw back, save the enormous costs and concentrate on its own backyard, leaving these countries to either turn to the Chinese and other importers of their oil and gas, or return to their former relative unimportance. Whilst the US, as the world's leading political and military power, will continue to focus on the stability of these regimes and engage here as elsewhere in great power rivalry with Russia and China, the urgency has softened.

Energy independence does change the nature of the game. And it comes with an added twist, again because of shale. It is the US that has the swing production now. Shale developments are very different from conventional large oil and gas projects. They can be turned on and off quickly and ramped up fast. Thus, as and when the oil price rises to the critical cost thresholds, back come the shale rigs. The US has both the prospect of energy independence *and* the ability to cap the upside on prices should OPEC try to flex its muscles and fix the market as it did in the 1970s and 2000s. Saudi Arabia, by contrast, no longer has the ability to swing production in the way it has in the past. As we shall see, its state finances are now precarious and oil revenues are needed to buy off its restless young population. It can no longer afford a big and sustained cut in production, and if it did, Iran and Iraq would probably fill the gap. Its attempts in 2017 to rig the market in collusion with Russia have been a failure.

An example of the US's new diplomatic freedom was the significant nuclear agreement with Iran in 2015, part of Obama's process of resetting the US–Iranian relationship. Prior to the revolution in 1979, Iran had been a long-term US ally, and before that a British ally. The Shah had been very much in the Western camp. Whilst many may see the force of sanctions as the reason for the agreement, much had been going on for over a decade behind the scenes. Iran is the main beneficiary of the fall of Saddam Hussein and the leading player in the fight against ISIS. Iran and the US found themselves on the same side in Iraq, though on opposing sides in respect of Israel and Syria, where Iran supports Hezbollah and the Syrian president. It is not completely inconceivable – especially under a post-Trump presidency – that, over the coming decade, Iran may begin to displace Saudi Arabia as the principal Middle Eastern US ally – a possibility that threatens the Saudis. The US can now afford to offend the Saudis. The balance of power has decisively changed.

The gradual US disengagement from the Middle East since the Bush presidency reflects US foreign policy 'weakness', but the coming of shale would probably have produced the same result in due course anyway. The shale revolution is a political bonus for the US, creating many more degrees of freedom for its foreign and military policies. Obama had a freedom that Nixon and Carter could only have dreamt about. Trump and his successors have even more.

The concerns will be more about rogue and failed states, especially if they become nuclear. These countries may end up mattering in the way North Korea does in its threat to global peace and security, but not because they have something the US needs. The capacity for the Middle East to cause global trouble is almost limitless. The Gulf states have rapidly rising populations, most are internally unstable, and several are, or are trying to be, nuclear powers. They are mostly prime examples of the resource curse: riddled with corruption, non-transparency, human rights abuses and the suppression of women's rights. Then there is the intractable Israeli–Palestinian conflict.

The pivoting of US foreign policy towards Asia, the containment of China and the maintenance of an open South China Sea are altogether more immediate in terms of military threats and challenges. Nixon may have gone to China, but it was a China that had little economic or military clout back in the 1970s under Mao Zedong's ruthless regime. That has changed dramatically, as China tests Japan's willingness to defend itself and militarizes the South China Sea. Russia is not the only revisionist state. China is playing this game too. Trump has largely followed Obama's pivot.

What has changed is the rationale. The US's interests are all about world order and stability, and its role as the usually reluctant world policeman. Its history is punctuated with bouts of isolationism, ended by wars caused by others which it finds impossible to keep out of. This was true of the First and Second World Wars, into which US politicians dragged their reluctant population. The US has never had an empire to defend, and even in its own backyard – in Mexico, the Caribbean and Central America – it has proved reluctant to impose its will through overt military force.

The recent exceptions under Presidents Ronald Reagan and George W. Bush should be considered in the context of this enduring reluctance. Nicaragua, Grenada, Iraq (twice) and Afghanistan were all wars with the exit very much in mind. None led to permanent control and colonization, in contrast to Putin's territory-based aggression. There are no South Ossetia, Abkhazia, Crimea and eastern Ukraine examples on the US shopping list. What the US wants from the Middle East are stable regimes with an uninterrupted flow of oil and gas to its friends and allies.

Stability for the US comes in various shapes and sizes. At one level it is about controlling the spread of nuclear weapons, keeping out Russian interference and tempering China's global outreach – the global balance of power between the remaining big players. At another level it is about more direct threats, particularly terrorism. The terrorist threat haunts the US. Unlike Europe, it has never been invaded. Wars on US soil have been about independence (from the British) and territorial integrity (the Civil War between the North and the South). It never suffered from the Palestine Liberation Organization (PLO) in the 1970s in the way the French and Germans did. The 9/11 attacks in 2001 forced the US to recognize that threats come not just in military hardware and from known major enemies, but also in a much more diffuse and amorphous form. There were no longer obvious targets to go out and get.

The terrorist menace emerging from the Arab and Persian worlds could not be contained in the ways that the generals understood. It was deeply ideological, and involved a willingness to die for its many causes and a brutality which was altogether different. The oil curse, and the Wahhabi–Saud relationship, has left a legacy that will take decades to work out. Furthermore, as the oil price fails to keep up with expectations, the ensuing economic crises in the Middle East will probably make matters worse. The US, therefore, cannot follow the full logic of energy independence towards complete disengagement, but it can at least avoid the direct economic consequences that did so much damage to its economy in the 1970s. It now has much less to lose.

In addition to the US interests in limiting terrorism and encouraging political stability in the Middle East, a third level of engagement concerns trade. Outside oil and gas the Middle East is significantly concerned with arms deals.[8] But military sales do not matter much in the bigger scheme of things. Despite Trump's obsession with trade, its share in the US economy is actually quite small. Most US trade happens between its states, on the inside, and not on the outside. Even where the US does trade externally there are important changes underway, and not just in the reduced role of oil and gas imports. The great offshoring of manufacturing (all those iPhones made in China) has probably passed its peak. Abundant, cheap domestic gas has already changed the petrochemical and refinery maps, and this cheap domestic gas has fed through into an improvement in wider industrial competitiveness. With greater energy independence, trouble in the Middle East spilling over into oil markets might actually improve the US's relative competitive position. Its worries here are more about the damage to its allies. But even here our second and third predictable surprises will gradually weaken this grip too.

Although these various overlapping factors in US foreign policy will take time to play out, and will be tempered by Trump's 'America first' priorities, historians will probably write up the current decade as the one where the US broke free from the shackles created by its external energy dependencies. The importance of this cannot be underestimated. The Middle East has absorbed much American time, money and blood. It has aligned the US with some very nasty regimes and there is little to show for all these sacrifices. It now has the option to turn to other critical challenges, like China and the South China Sea.

Win two: the carbon constraint – the second US surprise

Win two for the US relates to our second predictable surprise. In the conventional green NGO narrative, the US is often regarded as the great environmental villain. On top of its addiction to oil, and the great threat that it was supposed to face from running out of the ever-more-expensive supplies, the US has, until recently, appeared to be challenged by its addiction to carbon. Climate change, it has been argued, would not only do a lot of environmental damage to the US, but would prove an expensive Achilles heel in terms of the costs. The Republican Party in particular contained a rogues' gallery of climate change 'deniers', with George W. Bush and Dick Cheney having been especially prominent. Trump has now joined the list.

Recall how the narrative went in the 2000s, especially as promoted by the Europeans. The price of oil (and other fossil fuels) was confidently predicted to go ever upwards. The Europeans (and especially the Germans) would, by going for renewables early, be insulated from the price and supply shocks to come, and wind farms and solar panels would accordingly provide a cheaper alternative. Europe would have relatively cheap energy in the peak oil world, and hence US industry would suffer a big loss of competitiveness. Energy-intensive companies would choose Europe over the US. Wise Merkel and Barroso would triumph over foolish Bush and Cheney, not just on the moral high ground of mitigating climate change, but in the global economic competitive challenge too.

As already noted, and as further explained in Chapter 8, it is the Europeans that now look foolish. The oil price has not followed their preferred path; nor is it likely to. European energy-intensive industries have been decimated. The US, by contrast, emerged from the Great Recession with stronger growth and employment than Europe, and its fossil fuel dependency has turned out to be a big competitive advantage.

Yet the US is not immune to the global political pressure to be seen to be doing something about climate change, and it is not immune to the new markets and technologies that the push towards low-carbon energy will bring. It has therefore begun to develop its own decarbonizing policies, even as Trump has tried to reverse them.

Much of this has been in the rhetoric. Obama did a deal with China ahead of the Paris COP, and claims to be 'saving the planet' as a result of the credit he claims for the Paris Agreement (terminology he duly repeated when the US and China formally ratified the deal nine months later). Much of this is hot air: as we saw in Chapter 2, Paris is more the end of the road for this type of global top-down diplomacy, and its numbers do not add up. Trump's exit actually makes little difference.

Fortunately for the US, the Paris failures are less important than what is happening on the ground. In line with the global averages (but way below China and India), the US has been generating around 40% of its electricity from coal. But here the similarity ends. US exceptionalism in terms of climate change arises from the availability of plentiful and secure supplies of gas. The US can decarbonize quite quickly by switching from coal to gas. That indeed is what it has been doing, and that is why its emissions record stands up well against that of Europe, as described in Chapter 2. Regardless of whether anti-coal legislation stands the test of the US legal system, and whether Trump succeeds in his pro-coal approach, the opportunity is there anyway, and few will contemplate building new coal power stations in the US given the competitive alternative of abundant shale gas. That is one of the reasons why coal companies have been going bust.

The US has the coal-to-gas option on a much bigger scale than the other major players. Europe depends on imported gas, and much of it from Russia. China imports more expensive LNG, as does Japan and India. This gets the US over a big chunk of the transitional carbon period at almost no cost to its industry. Few other countries are in this lucky position.

Its addiction to the car and the internal combustion engine is a tougher challenge. As we have seen, the car explains much of the growth of oil demand in the US, and the US is a country that lends itself to the car – often of the bigger, SUV variety. Americans are price-responsive: as the oil price has fallen the demand for larger, less fuel-efficient cars has risen. The US policy approach to cars, like its current approach to coal, is to avoid the price mechanism. Ironically for such a capitalist country, carbon and petrol

taxes are neglected in favour of regulation. In the case of cars, these are the
CAFE rules referred to above.[9] Regulation works, even if some companies,
like Volkswagen, try to cheat. Over the last two decades it has had big
impacts on efficiency. The outcome is that the US carbon record is at
least as good as the European one, despite much lower taxes and the
greater scale of deindustrialization in Europe. Looking ahead, there is much
greater scope at least at the margin to make more progress in the US than
in Europe.

Win three: the US and the new technologies

For cars and in electricity generation, the argument made in Chapter 3 is
that it is the new technologies that will really make the difference. Here the
US holds most of the cards, and it is most likely to be a winner. Almost
everything that is required is in place. From strong R&D in its universities,
military and research labs, through to a corporate culture of entrepreneurship, its connectivity to financial institutions (and tolerance of bankruptcy),
and the openness of its intellectual and political systems, the US has what it
takes. Europe has much of this potential too, as we shall see, but China's
authoritarian regime has only state-driven corporations, censorship and
state finance. The US is the place where new technologies are most likely to
flourish.

It is not hard to describe these advantages. Starting with the great
research institutions, it was the military that developed nuclear technology,
email and the Internet. It is here that robotics and AI are likely to prosper,
as warfare goes digital and intelligent. Though the technical opportunities
gained by having a large military are shared with Russia and China, neither
of them has the military *in combination with* the other advantages that the
US holds. MIT and Stanford and other great universities have repeatedly
trotted out Nobel Prize winners (only the UK rivals on a per capita basis).
Its corporates have overwhelmingly dominated the communications technologies – again and again. AT&T and IBM once managed this dominance,
then came the new wave which gave us Microsoft, Apple, Google and
Facebook. The Europeans, with their Nokia and Siemens, are also-rans in
this game, and there are no new major British players. China can copy and
imitate, but not lead. Only Japan and South Korea are other serious
contenders in this technological race.

Looking ahead, it is not hard to envisage the new US giants of 3D printing, robotics, AI, batteries and electric cars, and solar generation. It is in the culture, and the US leads particularly in software and robotics. The established players are all on the front foot. New disrupters like Tesla are forcing the conventional car companies into the electric space. The now mainstream companies such as Google and Apple are getting into mobility and driverless cars. And so on. There are some European players in the electric transport game, notably among the car companies, such as BMW and VW. But they are notable because they are exceptions. China, India, Japan and Korea have players trying to get into this space, but outside the US only Nissan and Toyota look like potential winners so far.

Out in front

The three predictable surprises set out in Part One jointly create massive opportunities for the US. It is potentially the world's winner. The oil independency that shale has made possible, and the lower prices, takes away what was seen in the 1970s as an almost existential threat to the US's economic prowess. Now this particular economic threat has gone away, even if the political one that terrorism brings has not.

What the second and third predictable surprises bring is the prospect of the oil problem further diminishing as the century unfolds. The US is not just weaning itself off Middle Eastern dependency; it is also, like the rest of the world, gradually weaning itself off oil in general.

Watching the US struggle with climate change, and in particular climate change policy, may give rise to scepticism about its ability to decarbonize. Congressional reluctance to endorse Kyoto, and resistance to the suggestion of legally enforceable targets at Paris, indicate that many in the US see climate mitigation as more of a threat than an opportunity. Trump is doing no more than his predecessors, even Obama, who presided over the great expansion of fossil fuels. But this is to confuse two things: the politics of the vested interests of those who would lose out from decarbonization; and the economics of technical change and the opportunities for the US to develop new industries.

CHAPTER 5

The Middle East
MORE TROUBLE TO COME

Even with prices at $100 per barrel, the Middle Eastern oil producers were struggling to keep a lid on their populations and to stop all-out war between the major players. The Arab Spring brought down the regime in Tunisia, led to revolution in Libya and a continuing civil war, caused the overthrow of the military in Egypt before it resurfaced in a new guise, and kicked off the civil war in Syria, which in turn played into the eruption of ISIS in Northern Iraq and parts of Syria itself. As Syria descended into chaos, it brought Iran into a proxy war with the Arab Gulf states, and brought the US and Russia into the conflict on opposite sides. Then a proxy war between Iran and Saudi Arabia broke out in Yemen. With unrest in Bahrain, and the continuing Arab–Israeli conflict, it is hard to find any peaceful parts of the Middle East. To this can be added the continuing low-level war between Turkey and the Kurds. And almost all of this *before* the falls in oil prices.

This turmoil is not new. Conflict has been endemic since the break-up of the Ottoman Empire at the end of the First World War, with oil simply adding fuel to the fire. The Balfour Declaration in 1917 cleared the way for a Jewish homeland in Palestine. There followed the Sykes–Picot carve-up creating Iraq and Syria;[1] the repeated French interventions in Lebanon; the emergence of Ibn Saud from the southern deserts; the Iranian coup d'état in 1953 (with the help of the CIA); the Suez Crisis in 1956; the short-lived British-supported Iraqi monarchy; and then the coming of Saddam Hussein. Since the 1970s we have had the Iranian Revolution in 1979, the Iran–Iraq war in 1980–88, the Iraqi invasion of Kuwait in 1990, and then the US-led coalition invading and overthrowing Saddam Hussein in 2003.

The causes of many of these conflicts pre-date oil, and some go back as far as the early succession to the leadership of Islam after the death of the Prophet Muhammad. But what has made the Middle East important in world affairs is oil. Oil has exacerbated the underlying tensions, it has been a resource curse, and it has drawn in the great powers – again and again. Now that the oil price has fallen back, and with the prospect of a medium-to-long-term decline, the Middle Eastern producers are for the first time confronted with trying to work out how to escape this terrible resource curse-in-reverse and steer themselves to a post-fossil fuel world. It is a big ask.

To understand the nature of the problems they confront, and what the lower oil prices in the medium and longer terms might mean, it is first neces- sary to understand how they got to their current perilous state. The popular narrative is all about OPEC, about the central role of Saudi Arabia as the swing producer, and about the current price falls as part of a clever strategic game played by the Saudis. It assumes that these countries still have the power to move the price, and hence assumes that, eventually, OPEC will restore order and return to the good days of ever-higher prices. This narrative is profoundly wrong, and the consequences for the Middle East of the new reality have barely begun to be felt, let alone understood.

Pre-OPEC: the long struggle for a fair share of the oil revenues

At the beginning of the twentieth century, it gradually sank in for the major Western powers that oil would become the key resource to power their military forces. The scramble to secure supplies was a priority in the run-up to the First World War. The US had its own supplies, as did Russia, but the remaining global power, Britain, had none. Nor did France or Germany. Controlling the Gulf was already an imperative for Britain given its over- arching need to secure the trade and military routes to India. Churchill's decision to convert the Royal Navy to oil made this control a greater neces- sity, given that Iran provided the early supplies.

In the immediate post-First World War context, the Middle Eastern producers were not regarded as powerful countries to bargain with. They were more like colonies, and in exchange for supporting local elites, kings and dictators, the Western countries struck deals that left most of the economic rents with the oil companies (which were mainly proxies for their sponsoring national governments). The history of these relationships is one long struggle by the Middle Eastern countries to get a better deal. This would play out in the 1950s in Iran and then later in Iraq and Saudi Arabia, as the scale of reserves and the cheapness of extracting them emerged.

The early days were all about a small number of major oil companies, which carved up the Middle East between themselves. For more than half a century, these companies effectively controlled Middle Eastern oil production, and in doing so determined the contract terms and managed the international price. Implicit and often explicit collusion and cartels prevailed.

The deals started before the First World War, and before the Ottomans sided with Germany. The Turkish Petroleum Company (TPC), set up in 1912 to exploit Ottoman resources, brought together Shell, Deutsche Bank and the Turkish National Bank. The Great War kicked off a process of rearranging the chairs. The Anglo-Persian Oil Company (later BP) took over the Turkish National Bank's shares in 1914. Deutsche Bank was expropriated during the war and its share was eventually given to the French at the San Remo Conference in 1920, which decided the fate of the former Ottoman territories (and incorporated the Balfour Declaration). The US was late to the table, partly because it had its own oil back home, and partly because it was not a colonial power in the Middle East.

In the tussle over the Ottoman territories and access to their oil, the US was explicitly excluded from San Remo, afterwards retaliating with its own Mineral Leasing Act in 1920, covering the authorization of the oil, gas and coal rights on public lands. This meant the US could exclude companies from access to US resources, and as a result the US was invited back into what by then had become the Red Line Agreement of 1928. Thus began the great US entanglement in the Middle East. The parties agreed that they would not act independently in the TPC areas. Companies, and the governments behind them, fixed the oil industry in the Middle East, and this would remain the case right through to the formation of OPEC in 1960 and beyond.

The growing dependency on Middle Eastern oil worked both ways. The producing states grew to rely on the revenues to underpin their autocratic and often brutally repressive regimes, to provide a basis for their various attempts at modernization, and to help them cope with population growth. The consumer countries developed oil-based economies. Both sides became increasingly dependent on each other, and the relationship required careful political as well as economic management. The balance was maintained through military power, diplomacy, bribes and corruption, and all the panoply of ways in which such relationships between resource-rich and resource-hungry economies have evolved.

As demand grew, it was inevitable that the producers would seek revisions to the contracts they had had little choice in signing in the early years. From the perspective of the oil producers, they were being exploited by neo-colonialists (and they were broadly right to think this). Although there is a natural tendency to equate OPEC with the Middle East, nowhere was this felt more viscerally than in the early 'petro-states', Venezuela and Mexico. As an early large-scale producer, Venezuela was, with Saudi Arabia, a driving force behind the creation of OPEC.[2] The fight was played out through an endless and permanent process of contract renegotiation, punctuated by the occasional violent and revolutionary responses.

The producers had little leverage prior to the Second World War, but thereafter the balance of power gradually shifted. The war had reduced the capacity of Britain and France to retain a political, military and economic grip on these regimes, whilst the Cold War gave the producers other options, notably Russian ones. Britain and France were humiliated over Suez in 1956, and Egypt's President Nasser peddled his Pan-Arab nationalism on the streets, threatening the stability of Saudi Arabia, Iraq and Iran. Nasser's Egypt, and the United Arab Republic (the merger of Egypt and Syria, 1958–61) that Nasser promoted, was a direct threat. The Western countries now had competitors in what they had regarded as an almost exclusive zone of influence and control in the Middle East.

The US, with no colonial territory or ambitions, had its own backyard to contend with. The earliest to break ranks was Mexico, the world's second-largest producer in the 1920s. It nationalized its oil industry in 1938 to create Pemex. After the Second World War, the Middle East followed the example, starting with Iran and its ill-fated attempt to nationalize its oil industry in 1951. As noted earlier, the revolutionary government of Mohammad Mosaddegh was effectively overthrown by the CIA in the 1953 US–UK-led Operation Ajax. The industry responded to a US initiative and formed the 'Consortium for Iran' to share out Iranian oil production, reducing the share of BP to 40%.[3] The membership was subsequently named the Seven Sisters. It included four of the Aramco partners who controlled the Saudi Arabian oil industry, plus Shell, BP and Gulf. From then on the restored Shah followed a much more compliant policy with the Seven Sisters, who would continue to dominate not just Iranian oil but the wider Middle Eastern and hence world oil market, up to the first OPEC oil shock, with over 80% of the total market.

Notwithstanding the Iranian setback, the vicious and bloody Algerian rebellion against French colonial rule in the late 1950s and 1960s pushed forward the case for outright nationalization in the Middle East. After the Algerian nationalization led to the creation of Sonatrach, the next big impetus came in 1973 with the contract renegotiations and the so-called Participation Agreements, with the whip now clearly in OPEC's hands. The stage was set for a gradual process, replacing the Seven Sisters with a series of state-owned companies, dominated by Saudi Aramco, and supported by ADNOC in the United Arab Emirates (UAE), Qatar Petroleum and Kuwait Petroleum Corporation. Revolution in Iran in 1979 facilitated the completion of Mosaddegh's unfinished business from the 1950s. The Seven Sisters were replaced by the new state-owned companies, and these became the mainstay of the governing regimes' income.

Sheltering behind the politics of their autocratic governments, these companies had a capacity for secrecy and manipulation of their contracts and accounts that turned the problem of oil dependency into a deeper resource curse, weakening institutions and making corruption even more a way of life. The combination of secrecy and the fact that these enterprises employed comparatively few people spawned today's concentration of power and money in the hands of the (very) few.[4]

By 1980, these new nationalized companies dominated the Middle Eastern oil and gas industries, and were among the largest oil and gas companies in the world. The Seven Sisters now had to look elsewhere to book reserves, and this, combined with the new much higher prices, proved a major incentive to go after more expensive resources in Alaska, the North Sea, Russia and Africa. Without the OPEC price shocks in the 1970s, these developments would probably never have happened.

OPEC: never really a credible cartel

OPEC was the institutional format within which, and against which, this rebalancing of the relationships between producers, consumers and the old colonial powers played out. The early steps in the formation of OPEC focused on price, not output. It was set up as a result of a conference in Baghdad in September 1960 with Iran, Iraq, Kuwait, Saudi Arabia and Venezuela. Qatar, Indonesia and Libya joined soon afterwards, followed in the late 1960s and early 1970s by UAE, Algeria, Nigeria and Ecuador, and later still by Gabon and Angola.

With the hindsight of the oil shocks in the 1970s, OPEC came to be regarded as the textbook example of a cartel. The oil shocks created the myth of global market power, and hence the need for the world's consumers not only to take heed, but to try to manage that power. But in practice neither the producers nor the consumers could effectively collude, except for very short periods, and even then largely ineffectively.[5]

An effective cartel needs several conditions to be met: the product needs to be homogeneous; production needs to be controlled; and cheating needs to be detected and effectively punished. The benefits of the cartel price need to be shared out.

Put this way, it is obvious that OPEC fails most if not all of these tests. The key to maintaining a cartel is not only an agreement about how to share out production, but also how to prevent the obvious incentives to free-ride and thereby undermine any output-reducing agreement. These free-riding incentives are overwhelming: if others cut back and the price goes up, the incentive to pump as much as possible is very great. Add in political and regional rivalry, and a good dose of secrecy, and the wonder is that OPEC kept its game together at all, and why it took until the mid-1980s for the price shocks of the 1970s to dissipate.

There have been repeated attempts to create a sense of common Arab purpose, which could in turn reduce the incentives to free-ride. Arab nationalism as promoted by Nasser in Egypt in the 1950s and 1960s is one example, building on an early collective sense of purpose under the Ottomans. Common cause is, however, predicated on a shared under-standing of the political space. For the Arabs, the secularists in Egypt and in Turkey were up against the clerics. Its Wahhabi fundamentalist founders define Saudi Arabia. Then there is the fact that Iranians are not Arabs. Within the clerics, the Sunni–Shia split reflects the political Saudi–Iran division.

Little of this is surprising. Very few areas of the world display homoge-neity of ethnic origin or religion, and the Middle East is no different. Europe is rife with religious heterogeneity around the core split of Catholics and Protestants, and has many different ethnicities. The Europeans find it hard even for a subset of core members to fix their currencies. Fixing major sources of national income like oil is to the Middle Eastern countries an altogether bigger deal. Religious and ethnic divisions in the Middle East are, if anything, deeper now than at any time in the last century, and the melting pot of Christians, Jews, Sunnis and Shias has become murderous.

The prospect of reaching a common political understanding – a necessary condition for the cartel to succeed – is poor. Arguably this has occurred only once, in the early 1970s, kicking off an upward path to prices – the first great aberration in the long-run oil prices described in Chapter 1 – and hence a continuous rise in incomes in response to tightening production. Following an earlier attempt at an embargo in the 1967 Six-Day War, the catalyst was the Yom Kippur War, which provoked the temporary sense of Arab unity that had eluded Nasser in the 1950s (though, as noted above, he did manage to merge with Syria to create the United Arab Republic briefly at the end of the 1950s).

The coordinated attacks on Israel by Egypt and Syria provided a temporary sense of common purpose. Caught unawares, this provoked a sharp response from Israel and, despite its tiny population, it proved more than a match for the Arabs. As the US and other countries resupplied Israel's military, OPEC came into its own, imposing an oil embargo.

Israel and the Yom Kippur War may have provided the context, and may have encouraged an element of Arab unity, but it was more a pretext than an explanation, otherwise prices would have fallen back sharply once peace was restored. Indeed, the earlier Six-Day War in 1967, during which Israel grabbed parts of Sinai, the Gaza Strip, bits of the West Bank and the Golan Heights in a stunning display of military strategy, did not produce the response that Yom Kippur did.

Israel's very existence has been a complicating factor in the Middle East since the Balfour Declaration and the creation of a homeland for the Jews in Palestine in 1917. The Israeli problem for the Arabs has been around for most of the history of the Middle Eastern oil industry, but it had had little impact on oil supplies or prices. In 1967, OPEC did try to implement an oil embargo. It lasted a few days before it collapsed. The attempted oil embargo in 1973 was altogether more effective. The oil price increase stuck, and it caused severe economic pain to the US, Europe and Japan.

It was a new bonanza for the OPEC countries. Suddenly the yokes of colonialism had been dramatically thrown off. Money poured in faster than their economies could absorb it, creating major financial flows of 'hot money'. Given the lack of alternative oil supplies, it seemed as if the new price levels were going to stick for the indefinite future. Cutting production produced stunning results. Collusion proved attractive. Indeed, when it came to 1979 and the Iranian Revolution, if anything it looked as if OPEC had not been ambitious enough in the mid-1970s.

The gradual decay of support for the Shah of Iran is more apparent with hindsight than it was at the time. Iran had been a staunch ally of the US in a geographically sensitive area, bordering the Soviet Union to the north and Afghanistan to the east. After the 1950s attempt at nationalization, the Shah's line on Russia was altogether more helpful than that of Nasser in Egypt (until President Sadat's time), and helped to keep the Russians out of the Gulf.

When the Shah's regime collapsed, few could have anticipated just how radical and clerical the revolution would turn out to be, and how rabidly anti-American. The hostage crisis referred to in Chapter 4 rocked the US, and effectively destroyed what was left of Carter's presidency. Suddenly Iran was no longer supplying oil to the West, and an already highly volatile oil market responded. Prices doubled, to peak at $39 per barrel – equivalent to almost $140 today – a price reached again only briefly in 2008. To illustrate how extreme this 1979 price was, the January 2016 price ($27 a barrel) was even less *in nominal terms* than it was in 1979.

The high prices contained the seeds of their own destruction, as they would do again in the run-up to 2014. After the 1979 doubling in oil prices, it was assumed that the future path would be ever upwards. The OPEC countries could now squander their fabulous newfound wealth on the assumption that there would be plenty more to come. So when the prices eventually fell back by 1986, many OPEC members were left with high spending and much lower revenues.

The immediate cause of the mid-1980s price collapse was the decision by Saudi Arabia to increase output. As described in Chapter 1, the Saudis had borne the brunt of the cuts after 1979, and they opened up their production in retaliation. It is all eerily familiar: the same broad strategy has been deployed since the end of 2014.

As the price falls became entrenched in the late 1980s, some OPEC members suffered more than others. Kuwait with its tiny population could get by with lower revenues, and the Saudis were playing a long game. Iraq, on the other hand, suffered significantly and it agitated, to no effect, to get the Saudis to cut back production. The war with Iran had decimated its revenues.

Frustrated by the lack of response, and invoking its long-running (and not entirely spurious) dispute about borders, Iraq then took the fateful decision to invade Kuwait and set fire to its oil wells. The invasion caused local panic, including in Saudi Arabia, which feared that it was the real target.

The First Gulf War, as it became known, led to a temporary price spike, but once hostilities ceased, prices fell back for the rest of the 1990s, drifting down to $10 by 1999, and creating a new conventional wisdom that prices would stay low for the foreseeable future. *The Economist* even ran a story questioning whether oil companies could weather $5.[6] Allowing for infla-tion, $10 in 1999 was equivalent to the price back in 1970 – as if the whole OPEC-driven decade of the 1970s had never happened. The Middle East no longer looked so threatening. The 1970s looked like an aberration (which it was), and the very long-run gradual fall in prices reasserted itself.

But low prices, then as now, had their effect. Exploration and produc-tion (E&P) had been scaled back in the name of cost-cutting, and demand responded. In the early 2000s, the market tightened – not through anything OPEC did, but rather in response to market forces. The global stock market collapses in 2000 triggered a sharp fall in global interest rates, and an asset-driven boom got underway in the US as George W. Bush cut taxes and thereby loosened fiscal policy. All the while, the dramatic growth of China and its appetite for not just oil but the other minerals as well was gaining momentum.

The gradual nature of the price rises in the early 2000s was remarkable given the concurrent political turmoil. In 2001, 9/11 shattered illusions about the stability of Saudi–US relations, and Al-Qaeda set the pace for the new US 'war on terror'. There followed an ever-expanding cycle of violence and extremism. The US retaliated (with allies) with the bombing and inva-sion of Afghanistan and the invasion of Iraq. What started with the planes flying into the Twin Towers would morph into an Iraqi civil war, which continued through the next decade. Sectarian violence inside Iraq escalated as the US in effect pushed Iraq into the Iranian sphere of influence. The Arab Spring, the fall of Gaddafi in Libya and Hosni Mubarak in Egypt (and the temporary rule of the Muslim Brotherhood), the Syrian civil war, the emergence of ISIS in rebel-held Syria and then in Iraq, and the renewed Kurdish–Turkish conflict – all were part of the sequel.

Saudi's responses

Through these wars the oil price gradually rose, not even dented by the credit crunch and financial crisis in 2007–08. The oil price carried on going up, peaking at $147 in July 2008. There was a short, sharp collapse in December 2008 to touch almost $30, and then the price climbs resumed. With Libya and Iran out of the market, through civil war and sanctions respectively, and Iraq's oil industry still crippled, Saudi Arabia showed no inclination to come to the rescue. All the while, China's dramatic economic growth tightened the demand side. The problem in the late 2000s was not one of OPEC countries exerting their power, but rather that they could not get their production act together. This time, the second great aberration in the long-run oil prices, and as noted in Chapter 1, it was all about demand – the first time it was about supply.

By 2014, the OPEC countries had felt the full force of the resource curse. In response to the Arab Spring, and more generally working on the assumption of ever-higher prices, the main players had spent up to the oil price at $100. Even Saudi Arabia needed high prices to balance its budget. It is quite plausible that, notwithstanding the threat from Iran to Saudi dominance in the region, $100 could have been maintained for the rest of the decade through to at least 2020, had it not been for the twin impacts of shale in the US and the gradual slowdown in China.

But both happened, and US shale presented the Saudis with a major challenge. If shale added so much output, it would either force the OPEC countries to reduce their own production to compensate, and hence accommodate the greater US market share, or force down the price. Worse still, shale gas took the US out of the gas-importing business and, as a result, gas reserves in countries like Qatar would be vulnerable to stranding.

Saudi Arabia decided to hold up its output in the face of the US increase in output, in order to force down the price and force out shale production. In the process, it could also signal to Iran and Iraq that it would not accommodate greater OPEC quotas at its expense. The Saudi bet rested on shaky assumptions. It assumed that the shale costs were higher than conventionals;

it assumed that it alone had the swing production; and it assumed that sanctions would continue to hamper the recovery of Iran's output. None turned out to be true: US shale costs followed the price downwards; the very flexibility of the shale technology meant that it could also play a swing role; and Iran continued to make headway. Even if Saudi Arabia forced shale out, it could come back much quicker than conventional oil wells. And Iran and Iraq would not go away.

More Iranian and Iraqi oil to come

Having kicked off a price collapse, OPEC now faces the consequences. As with the mid-1980s oil price fall, it is very hard to re-establish discipline once broken. There are three big problems: the urgent need for the Middle Eastern countries to generate revenues to support their spending; the lack of any obvious constraint on production in Iran and Iraq; and the continued expansion of non-OPEC production, not just in the US but more widely as shale technologies start to be applied on a global scale, and sometimes in surprising ways (such as increasing output from existing wells).

The resource curse and the Arab Spring together drove up government spending to buy off opposition. The curse meant that much of the revenue was lost in corruption and inefficiency. The result is that the budget break-even price of oil that OPEC countries now need is much higher – in many cases around $100 per barrel. There were some who assumed that because they 'need' this higher price, OPEC would therefore restore the price to this level. This is a fundamental mistake: as the price falls, all the main Middle Eastern producers found themselves with large holes in their budgets. In order to survive, more revenue is needed to make up the shortfall. The paradox is that, at least initially, the lower the price the higher the output, not the other way round.

The international agreement with Iran and the P5 + 1 countries sets the scene for Iran to return to the global markets as sanctions are lifted, including access to financial markets and the technologies that independent oil companies can bring.[7] Iran has every incentive to drive up production in the short-to-medium term. It needs the money to balance its budget and to finance all its forays in Iraq (against ISIS), Syria (in defence of the Assad government and the Alawite minority), Lebanon (in support of Hezbollah) and Yemen (in support of the Houthis).

Iran has the need and it also has the potential. It added around 0.5 mbd almost immediately. Further out it could eventually produce as much as 10 mbd. Given the stand-off with Saudi Arabia, it is hard to imagine a scenario where it limits its production to fit an OPEC-wide initiative, and not before it has increased its output well beyond 5 mbd. Iran also has the world's largest reserves of gas to add to its oil potential, as well as possible routes north to rival Russian gas supplies to Europe.

Iran's potential augments that of Iraq, which has some of the lowest-cost production in the world. These reserves produced 4 mbd before Saddam Hussein and the almost continuous wars since 1980 undermined this potential. Yet, even with the scale of unrest, production has been creeping up again to over 4 mbd. In 2014 it was one of the leading contributors to global oil supply growth (after the massive US contribution, and is up there with Iran and Canada), as Figure 5.1 illustrates.

Iraq's reserves are concentrated in the south (around 75%) and the rest in the Kurdish north-east of the country. Although building up production

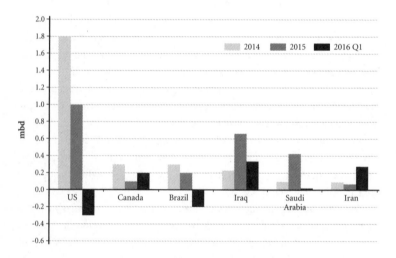

Figure 5.1 Top contributors to world crude oil production growth, 2014–16

Sources: OPEC, 'Annual Report 2014', Organization of Petroleum Exporting Countries, 2015, http://www.opec.org/opec_web/static_files_project/media/downloads/publications/Annual_Report_2014.pdf; OPEC, 'Monthly Oil Market Report', Organization of Petroleum Exporting Countries, 13 May 2016, http://www.opec.org/opec_web/static_files_project/media/downloads/publications/MOMR%20May%202016.pdf

Potential medium-term oil production

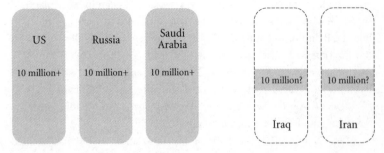

Figure 5.2 Oil production in the medium term

will take time – perhaps a decade – the medium-term potential is enormous. The reserves, already ranking around fifth in the world, are likely to rise as they are underexplored and the costs are among the world's cheapest, perhaps as low as $2 per barrel. There is as a result no upper production limit in sight, and Iraq could in theory surpass even Saudi Arabia.

Taken together, Iran and Iraq could eventually add another 10 mbd to world output, above and beyond their near-term prospective joint output of around 8–9 mbd, taking world supply comfortably to over 100 mbd. Iran and Iraq would then join the US, Russia and Saudi Arabia in the 10 mbd club. There would be three Middle Eastern countries in the game, two of which are overtly hostile to each other, and the third, Iraq, split between them, with Sunnis in the north and Shias in the south.

The consequences: Saudi Arabia's declining economic significance

The negative consequences for the Middle East of a gradual drifting down of oil prices and the eventual ending of the age of fossil fuels are hard to underestimate. Of all the Middle Eastern states, Saudi Arabia has arguably the most to lose. It might be tempting to conclude that having come out of the desert a century ago, Saudi Arabia might be about to go back again.[8] If it does, the consequences are unlikely to be a peaceful geopolitical unravelling of the great twentieth-century economic boom for the Middle East more generally.

The Saudis should be credited for seeing this threat coming.[9] It has after all been all too visible. Al-Qaeda, and Osama bin Laden in particular, had the Saudi monarchy in its sights. The Arab Spring led to uprisings in neighbouring Bahrain, and induced a big rise in government spending to buy off the discontented. The re-emergence of Iran now once again threatens Saudi regional hegemony. Saudi sponsorship of militants in Iraq, Syria and Pakistan have come back to haunt it, and the Saudi Wahhabi religious fanaticism that provides the underlying ideology is incompatible with the development of a modern industrialized country.

There is much for the bulk of the population to be discontented about. The Saudi royal family is an extensive one. Ibn Saud produced at least forty-five sons among his estimated hundred children, and had around twenty-two wives. This proclivity is reflected in the sheer size of the resultant royal family. They dominate the political and economic life of the country and they take most of the money. For the rest of the population they are one of a number of sources of resentment.

Saudi Arabia's population has grown from around 4 million in 1960 to 30 million in 2015. The median age is around eighteen. This is a trend set to continue: the total fertility rate remains over the critical level of two births per woman. As with Russia and China, autocracies have tried to buy acquiescence with economic growth. For Saudi Arabia and Russia this has meant oil revenue.

Any country faced with these demographics would be challenged, but when these are overlaid with the massive gulf between the royal family and the rest of the population, the religious fundamentalism and the fact that almost all of the money comes from oil, the sheer scale of the challenge emerges. It was already enormous with high oil prices; with lower ones, and the prospect of prices gradually declining in the medium-to-longer term, it would take a miracle to emerge from this unscathed.

Saudi Arabia has to maintain its tight grip on the population, maintain its relationship with its clerics, and defend itself from Iran and its proxies. Iran shares similar challenges: it relies on its clerics too, it has its own kleptomaniacs in the form of the Red Guard, and it needs the oil revenue. But Iran has advantages that Saudi does not. It has a much deeper economic history, a broader economic base, and a longer tradition of trading. Iran was important long before oil came along, whereas Saudi Arabia has been significant only since and because of oil. Indeed, before oil it did not exist.

The Saudi plan, 'Vision 30', is to diversify into other industries to create a modern economy.[10] At present, there is almost nothing the country produces that anyone else wants other than oil and gas. Indicative of the simplistic approach that comes of dictatorial monarchy, part of the 'solution' appears to be hiring consultants eager for the fees and anxious to sell the story. The 'plan' is to sell 5% of Saudi Aramco, create a new sovereign wealth fund with the proceeds, and then spend the money developing new industries that its consultants have selected for it, in the hope that it will be able to substitute for imports and build competitive advantage beyond oil.[11] It is as if economic growth is a matter of political will and can be driven by those in charge, and that expertise can be bought on the open market.

There is almost nothing in economic history to suggest that this sort of strategy will work, and lots of reasons for thinking it will fail. Whilst authoritarian leaders can create the conditions for transitional growth – as, for example, in China and Singapore – there needs to be a set of prior conditions in place to make it happen. Competition plays a key part, as do resources and export markets. Since the Second World War, all the great transitions – Japan, Germany, South Korea and China – have been based on exports *and* a welcoming importer in the shape of the US. They have all been based on manufacturing, with products attractive to the US markets, and with American (and to a lesser extent European) companies providing key technological skills. It is hard to think of any parallels between these examples and the desert kingdom. What could Saudi Arabia possibly produce better and more cheaply than others to compete in the big world markets? How could they match their current imports with domestic substitutes? How exactly an educational system based on religious instruction is supposed to deliver the necessary expertise and human capital to support these new 'winners' remains a mystery. It is therefore not surprising that Vision 30 has little to offer in terms of alternative exports to oil. It is largely confined to import substitution.

Saudi Arabia's attempts to develop new industries will cost money that would otherwise be spent on buying off its restive population. That opportunity was provided in the great commodity super-cycle, and it was not taken. Instead, a large budget deficit has emerged (already around 15% of GDP per annum in 2015), and even with major expenditure cuts, the link between the riyal and the dollar has come under speculative attack.

Saudi Arabia does have advantages that others in the Gulf lack. It has significant financial reserves, and its tax base is almost untouched. It can spend the money it has saved and it can introduce VAT and income taxes on foreigners in the first instance. Yet neither of these looks like a panacea. Despite the enormous oil revenues over the last decades, the financial reserves are actually rather small and are being depleted quite quickly. That is why it has had to contemplate selling a chunk of Saudi Aramco and starting to borrow in international capital markets already. Although the scope for tax increases is great, so too are the political consequences. The rich, essentially the royal family, would have to pay. In effect, the wealth of the few would have to be siphoned off to create new industries for the benefit of the many non-royal Saudis. It would take a remarkable degree of enlightened self-interest for this to happen, especially as the leading members of the family have already taken quite a lot of their wealth elsewhere – into Scottish castles, Swiss bank accounts and US property. That is why the easier option of selling off bits of Saudi Aramco has been chosen.

Many countries faced with these sorts of challenges end up relying on bailouts by global powers which are looking out for their client states and the geopolitical power this influence brings. Thus Russia has to bail out its client fringe, now including Crimea. China looks after a host of nasty and incompetent regimes, from North Korea to Zimbabwe. The US, as the remaining global superpower, looks after many Middle Eastern regimes. Although, unlike any of the above, Saudi Arabia does not yet need US aid, it does rely on it for military protection. This creates the paradox: the Wahhabis' 'Great Satan' is also their great protector; and Saudi religious fanatics can attack the US (as they did on 9/11). As we have seen, although this relationship has at times been fraught, especially when it comes to Israel's survival (as, for example, in the Six-Day War and the Yom Kippur War), it has nevertheless endured. It was the US that dealt with Saddam Hussein's invasion of Kuwait (and thus protected Saudi Arabia), and it is the US that provides the military hardware and support for Saudi Arabia's air and ground forces (Saudi Arabia spends 15% of GDP on defence).

For Saudi Arabia, the terrible truth that flows from the development of shale oil (and gas) is that the US is cutting the ties of this dependency. As we have seen, it does not need the Saudis anymore, and any Saudi embargo could be much more easily accommodated than in the 1970s. The Saudis no longer present a credible threat. A setback to shale oil and gas development from the lower prices will do little to alter this new situation. The result is that the US guarantee given by President Roosevelt to the Saudi king on his return from the Yalta Conference in 1945,[12] partly in exchange for the dominance of Aramco and US interests, is beginning to ring hollow, and the relationship is slowly deteriorating. Lots of little spats add up to a trend. The US continuing to bring pressure to bear on the 9/11 perpetrators; the Saudi view that ISIS is the consequence of the US ushering in an Iran-backed Shia government in Iraq; and the more overt US focus on Saudi Arabia's appalling human rights record and its suppression of women (at least until Trump came along) – these are some of the irritants which are no longer so carefully managed. US disdain for the more medieval aspects of the Wahhabi religion became overt during Obama's presidency.

The Saudi royal family therefore finds itself on its own, with much lower oil prices, threatened by some of its own population and Iran, and with enormous obstacles to diversification. The country has a large number of idle males with little to do and much to be resentful about (the women don't really count yet).

What might happen? The answers that other examples throw up are as follows: emigration of both the brightest and best educated *and* the destitute; and internal revolution followed by some form of military authoritarian dictatorship. These two have a tendency to interact: internal civil unrest causes migration. The potential for revolt comes from the Shia minority, effectively in a state of constant low-level revolt already, *and* more importantly from the non-royal Sunni population, some of whom were attracted to Osama bin Laden. The Shia minority may seek support and sanctuary from Iran, and Iran may seek to come to its defence. But for the rest, emigration looks to be one of the few options (other than joining in others' wars as mercenaries). Where would they go? The attractive option is of course Europe. The Syrian refugee crisis in Europe may turn out to be a precursor of more to come, especially when the displaced from the other Arab oil countries are added.

Iran's future, and that of Iraq and Syria

Iran looks rather different from Saudi Arabia. As noted, its economy has much greater depth and it is not entirely oil-dependent. Indeed, Iran has proved able to survive with limited oil revenues under recent sanctions; it has had its revolution; and even as the new normal oil prices disappoint in terms of revenue, Iran's current oil and gas revenue is so low that it can only go up.[13]

Iran has more than twice the population of Saudi Arabia, even if its fertility rate has already dropped significantly. In due course, it will probably suffer the ageing population problem, long before Saudi Arabia has to grapple with this. It too has widespread unemployment and underemployment, worsened by war followed by sanctions.

Yet the structure of its economy is very different. Iran has also considerably increased its territory, for it now effectively controls the southern part of Iraq, with its Shia majority. The other two parts of Iraq are under Kurdish control, with Sunni extremists in the north and north-west spilling over into Syria. Iran supports the dominant but minority Alawites on the economically more important western seaboard of the Mediterranean, and has the Russians as its allies there.

If – and it is a significant if – southern Iraq and Iran are considered as one Shia block, then they jointly hold more oil than Saudi Arabia. Iraq's southern fields have enormous potential to add to Iran's oil and gas. Although in both cases, expectations are very low, and oil revenues are currently down, Iran can expect a positive relative income flow for the next couple of decades, even at $30–$40 per barrel. Whereas Saudi Arabia is coming down in total oil revenue terms as prices fall back, Iran and Iraq are going up. This low base is also an advantage when it comes to internal opposition. Both countries, starting low, have considerable opportunities to improve the lot of their populations. Life should get better in both Iraq and Iran, on the basis that it is hard for it to get any worse. But for Saudi Arabia, and the expectations of its people, things are going to deteriorate. The changing directions of progress, rather than its level, drive politics and revolts.

Further north in Northern Iraq and Syria, it is hard to imagine a rapid return to peace and normalcy. Geography here matters: much of Northern Iraq is sparsely populated, and the difference between economic success and failure can be small. If it were not for the wider international spillover of ISIS terror, Iraq's fate would not register much on the global scale. Syria is different: it has a long and deep cultural, economic and political history stretching back before the Crusades, based on its western seaboard on the Mediterranean and its importance in trade.[14] Reconstruction of its great cities, such as Homs, Damascus and Aleppo, would create an economic opportunity, and it could recover remarkably quickly. Migration might then cease, and some Syrians might even return. The key point is that it is hard to get much worse.

Looming over all these countries is Turkey. The Ottomans dominated much of the Middle East for centuries, and all the modern Arab states emerged from the ruins of the Ottoman Empire during and after the First World War.[15] Persia (Iran) has in many respects been defined by its long struggles against the Ottomans, and its territories have ebbed and flowed, notably along its Iraqi and Kurdish borders, and in Anatolia. The modern Turkey that emerged under Mustafa Kemal Atatürk took a secular direction, but this has not proved enduring, and for the last decade Turkey has been taking a more Islamic (Sunni) turn. Since the end of the First World War, the Middle East has witnessed continued instability over the resulting boundaries. The missing ingredient has been the one group that lost out after 1918: the Kurds, who never got the state they demanded.

Taking an optimistic view of the northern Arab states, the post-Ottoman borders would probably have to be revisited, and in particular those of Iraq, Syria and the Kurds. The curse of the Sykes–Picot Agreement would also need to be revisited. Iraq might have to be split into three: Shia Iraq, Sunni Iraq and Kurdish Iraq. The Iraq–Syria border is mostly desert, and hence boundaries are not only less significant but mean less to the inhabitants of the regions too. But the Syrian Kurds in the north might form a band across the top of Syria and Iraq, right along the Turkish border. This in turn could further destabilize the large Kurdish population inside Turkey, and therefore raise questions about Turkey's own post-Ottoman borders.

Such speculation about possible political solutions immediately throws up massive obstacles, and makes the alternative pessimistic scenario more compelling. This is a state of continuous civil war – sometimes low-level violence, sometimes major battles and conflicts. It is not hard to imagine Northern Iraq and the Kurdish areas engaged in conflicts for another fifty years. From a geopolitical perspective, we are already living with the consequences, and have been for a century. The new challenges emerging are external to this blighted set of countries – mass migration, the export of terror and the risk of 'accidentally' triggering wars between superpowers. A new normal oil price, gradually declining in the medium-to-longer term, makes the likelihood of the pessimistic scenario much greater, since these countries are unlikely to be leaders in the new technologies discussed in Chapter 3, notwithstanding Saudi Arabia's Vision 30.

CHAPTER 6

Russia
BLIGHTED BY THE RESOURCE CURSE

Russia, as a fossil fuel economy, is a big loser from the fall in prices, and will struggle to come to terms with the consequences of medium-to-long-term declines. President Putin personally is exposed: he is the product of the collapse of oil prices in the late 1990s, without which it is doubtful that he would have progressed from being a mid-ranking KGB officer up to 1990, and a municipal bureaucrat in St Petersburg in the 1990s, to the presidency. He had the good luck to be chosen by Yeltsin just at the moment the oil price reached rock bottom and, as leader, he has benefited from the great commodity super-cycle.

Russia has few advantages in the new technologies, even though it does have a military capability. Its corruption, autocracy and the absence of leading global companies are not accidents: it is exactly what you would expect of a resource-cursed country. The collapse of its industrial base as the Soviet Union imploded as it faced international markets reinforced its dependency on oil and gas. Its dependency on the two commodities is greater now than it was under the Soviet Union. The contrast with the US could not be greater: the US is a clear winner, whilst Russia faces a bleak energy future.

This will all matter greatly for the Russian people. But it will have wider consequences too because Russia continues to have geopolitical ambitions, and it has lots of nuclear weapons. Despite being a declining power while China has been a rising one, both countries may well turn to nationalism and use 'external enemies' to bolster their domestic support in the face of new and much more difficult economic circumstances. The decline of oil revenues will not just be a threat to the internal stability of Russia, but it is likely to be a serious external threat too, exacerbating already difficult situations in Eastern Europe, the Caspian Basin, and further south into Syria and the Middle East. Indeed it already is, as the people of Syria and the remaining bulk of Ukraine have discovered.

Whilst the US has been vulnerable to oil prices too, and it has had its fair share of cartels, monopolies and corruption, as well as serious shocks from OPEC in the 1970s, and whilst it has chosen to go to war in oil-rich regions, oil has not defined the US in the way it has Russia. On the contrary, it has developed a diversified market economy in which oil is but one industry. Russia remains afflicted with the oil curse in a way that the US is not.

With this dependency, Russia is exposed to the full impacts of falling prices. The implications for its future global power and influence of a sustained lower price, followed by falls over the medium and longer terms, are not well understood. Russia stands exposed to the new abundance of oil and gas, to the decarbonization agenda and the host of new technologies. Without credible institutions it faces the very real prospect of losing what is left of the global status that Putin so obviously craves. Without the continually rising revenues that Putin benefited from in his first fourteen years in power, Russia may not be able to sustain its military build-up and the payouts to the many claimants on the Russian state. It could end up with a stranded political elite unable to buy its way out of trouble.

It is not hard to predict how Russia will react to its new comparative poverty. Its actions will be conditioned by its past history, dominated by natural resources, and the overwhelming dependency on European markets. The starting point is to understand its resource base and how this has developed, since this will determine its options. This sets the scene for Putin's more recent moves – in Ukraine and now Syria, and his opening to China.

Russia has always been a resource-based economy

Russia's dependency on oil and gas is not new, but stretches back into the late nineteenth century and the development of the Absheron Peninsula deposits near Baku (now the capital of Azerbaijan). There had been sporadic use of deposits seeping to the surface since ancient times, but it was not until the market was opened up to auctions in 1873 and the arrival of the Nobel Brothers that Russia began to rival the US in kerosene production. It has never escaped the grip of its fossil fuel industries, despite a number of attempts, notably in the early years of the twentieth century as the last Tsar tried to handle the combination of industrialization and rapid population growth, and later as Stalin attempted forced large-scale industrialization.

The Bolshevik Revolution in 1917 put an end to international cooperation and from then on Russia was on its own.[1] It had to pursue an autarchic economic policy within the confines of what was left of the Russian Empire, and it had to do this in a vast country with a small population. Most of the history of the Russian oil and gas industries is one of development in isolation, eventually colliding with Western and international companies and their technologies only after the collapse of the Soviet Union.[2]

After Baku came the Volga region, which sustained the Soviet oil industry from its first production in 1929 through to the 1960s when the big discoveries in Western Siberia – such as the Samotlor Field – gradually took over, on which Russia's modern oil industry is built. These fields were the mainstay of the industry and its exports right through to the current decade. Other developments, notably in Eastern Siberia, Sakhalin Island and in the Arctic, have added to the potential, but not yet surpassed those of Western Siberia, even as these decline.

When it comes to gas, Soviet developments depended on a small number of giant fields in Siberia (Medvezhe, Urengoy and Yamburg), and the development of pipelines to bring its output west into Europe in the 1970s and 1980s, primarily, though not exclusively, via Ukraine. It was only after 2000 that Putin turned to bypassing Ukraine as part of his struggle to regain control of what was once the core of Russia. Again, as with oil, the potential for new fields onshore and offshore is immense.

With both oil and gas, Russia is spoilt for choice as to where to look. There are lots of relatively underexplored areas; there are considerable gains from applying up-to-date technologies to the existing fields to improve the depletion rates; and then there is shale and, in particular, the large deposits in the Bazhenov Formation.

The outcome is that Russia has among the largest oil, gas and coal reserves in the world, and with these reserves it is hardly surprising that it has a very energy-intensive economy, and that it ranks among the largest exporters of all three fossil fuels, and especially gas.[3] It is an asset base whose value is greater today than in the future if oil and gas prices decline in the medium-to-longer terms.

The relationship between Russia and Europe has come on the back of the mismatch between Russia's abundant resources and Europe's lack of them. Russia has always supplied Europe with coal, but oil determined the modern relationships, before gas took over. The Hitler–Stalin pact of 1939 provided oil and raw materials to support Hitler's rapidly growing war machine, and indeed the failure to replace these supplies from Romania and elsewhere after Hitler invaded the Soviet Union undermined the German forces, notably in North Africa. This is why Germany (like Japan) ended up trying to manufacture synthetic fuels as the Second World War drew to a conclusion. Russia's appeal to Germany has always been the match between the need for raw materials and the export of manufactured goods and machinery, and it remains the driving force of their economic relationship today.

After the Second World War and up to the 1970s, the Soviet oil and gas industries grew in a context of stable global prices and abundant supplies. The industries were fully Soviet state-controlled and incorporated into the Soviet planning regime – *Gosplan*. When the oil price shot up in the 1970s, it was well placed to benefit, but in true resource-curse style it stagnated. It no longer had an urgent need to diversify. During the 1970s the Soviet economic growth rate (to the extent that it is a credible number) was not transformed. It may have had a more diversified industrial base then, but it was already falling behind.

Higher oil prices masked these trends and fooled many in the West into thinking that the Soviet Union was still an economic superpower. But it wasn't, and there was no growth boost from the high oil prices. What it did have was enough money to pay for its military adventures. By the start of the 1980s, the Soviet army was bogged down in Afghanistan, unable to tame its tribes and their resistance forces and, as the price fell back in the mid-1980s, stagnation threatened to turn into disintegration. Now it had neither modernized industries nor lots of oil and gas revenues to compensate. Whilst it was true that from the 1970s Europe was becoming more and more reliant on Soviet gas, and that the West Germans began to think seriously of improving relations with Russia itself, and with it some reconciliation with Poland and a thaw in the relations with East Germany as part of what was called *Ostpolitik*, the reality was that the Soviets were increasingly dependent on the European export markets.

The Brezhnev years are now regarded with the hindsight of the fall of the Berlin Wall and the implosion of the Soviet Union.[4] But these events were not widely foreseen at the time. Whilst Afghanistan turned into a nightmare for the Soviet army, and with Poland raising a challenge in Eastern Europe with the growth of the Solidarność (Solidarity) trade union in the Gdansk shipyards, few saw a collapse around the corner.

The immediate cause was the dominating feature of the period – the last twist in the Cold War. The election of Reagan in 1980 heralded a new US approach after the dismal Carter years. Reagan wanted to *win* the Cold War, not to manage it. The US increased military expenditure and the Strategic Defense Initiative (Star Wars, as it was known) threw down a gauntlet, which Russia struggled to afford to pick up. Reagan made it obvious to Gorbachev, Brezhnev's ultimate successor, that Russia could not keep up.[5]

The economic inefficiencies of the Soviet Union were the proximate cause, and the blame was laid at the door of socialist planning, Soviet style. But these inefficiencies were not new. Stalin's collectivization had undermined its agriculture, and his ruthless approach to internal dissent undermined its military forces, as Hitler demonstrated in the opening acts of his Operation Barbarossa campaign to conquer Russia in 1941. Russia is no stranger to both adversity and dictatorship.

What made the situation much worse in the early 1980s was that the 1970s had allowed the leadership to mask the real impacts of the resource curse. Brezhnev had spent the windfall from the 1970s, and to no obvious long-term benefit. When the pressure to react to Reagan's strategy with more military expenditure came up against the collapse of oil prices in the mid-1980s, the extent of the inefficiencies were there for all to see.

It is often claimed in the West that Gorbachev was a visionary leader who understood that socialism was ultimately unsustainable. Both Margaret Thatcher and Ronald Reagan thought he was a man they could 'do business with'. But it was much more complicated, and Gorbachev himself never really believed in markets (or indeed probably never even understood how they work); nor was he willing to accept the consequences of full democracy. Gorbachev thought he was rescuing the Soviet Union through reform, not burying it, and he strived right to the very end in 1991 to hold on to at least the core countries, notably Ukraine.

The sharp falls in oil prices in the mid-1980s helped to precipitate *glasnost* and *perestroika*, yet at the same time made it practically impossible

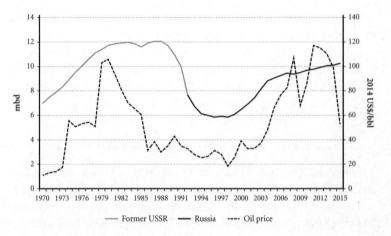

Figure 6.1 Soviet/Russian crude oil production, 1970–2015

Sources: EIA, 'Monthly Energy Review', May 2016, http://www.eia.gov/totalenergy/data/monthly/pdf/mer.pdf; BP, 'Statistical Review of World Energy', 64th edition, June 2015, bp.com/statistical review; Aurora Energy Research 2016, with data from Thomson Reuters

for the reform programme to be successful, even if Gorbachev had really wanted to move to a market economy. It was the stark exposure of Soviet economic weaknesses that paved the way for Gorbachev's initiatives, and at the same time left him short of the necessary financial cushion to facilitate the transition. It is still rarely appreciated that the fall in the oil price in the 1980s was a significant factor in what to contemporaries was a sudden and unexpected collapse of the Soviet Union.

The continued fall in oil prices in the 1990s left post-Soviet Russia in a precarious state. Yeltsin's attempts to build a functioning market economy and democracy were hopelessly undermined by the lack of money. This led to desperate measures to shore up his government, and he had to turn to the emerging oligarchs in the 'loans for shares' deal in 1996. This kept Yeltsin's political career alive and allowed him to hang on to the presidency, but at the expense of a massive asset transfer to a small number of businessmen. This is how much of Russia's natural resources passed into private hands.

Though there were many reforms under Yeltsin, Russia increasingly became identified as being under the partial control of a corrupting oligarchy which was widely hated. Putin got lucky, and the timing of his rise to the presidency was perfect. After the bankruptcy of 1998, the Russian economy had bottomed out. Things could only get better, and the oil price increased from the moment Putin came to power, allowing him to claim to be the saviour of Russia. As he regained the control that Yeltsin had lost, pensions and army wages could now get paid.

Regaining political control over the oil and gas industries

It fell to Putin to sort out the mess left by Yeltsin: to rein in the oligarchs, get control of most of Russia's oil and gas companies,[6] and use the revenues to suit his political purposes.[7] This should not have come as any surprise. He had already set out the case for controlling the commanding heights of the economy through central state control of natural resources in his doctoral thesis, written in his St Petersburg years.[8] This was not a blueprint, but it did set out a possible direction of travel. Putin's popularity in the early years of his presidency was in part due to his willingness to stand up to the oligarchs, though little did the population (or Western leaders like Bush and Blair[9]) understand that his aim was not to replace them with competition and markets, but to ensure that the key natural resources fell into his and his close associates' hands. It was not monopoly that Putin objected to, but rather others getting their hands on the economic rents.

Whilst Putin and his close advisers added other ideas, like the concept of the 'vertical of power', the 'dictatorship of the law' and the application of 'political technology' to flesh out his political philosophy, it was the control of resources that mattered most. Without the money, none of the others would even be relevant. The purpose of power as Putin saw it was to build a strong centralized state, and that state's power rested on its revenues. How this was to be achieved depended upon events. Putin has been a master of exploiting opportunities.

Having secured power in 2000 after launching a second war with the Chechens, Putin summoned the leading oligarchs to a meeting in the Kremlin in 2000 and spelled out the new rules of the game.[10] They could keep their money and some of their assets provided they kept out of politics. It was Putin, not them, who was now in charge. Otherwise, as the oligarch Mikhail Khodorkovsky was to discover, it would mean exile at best, or time in Russia's prisons followed by exile at worst.

The first real test of the new Putin doctrine was the Khodorkovsky case.[11] This mattered not just because of the power struggle between the two individuals – though the arbitrary use of state power in circumstances of mutual animosity played a role – but also because of the way in which Khodorkovsky built up the oil company, Yukos, and what he planned to do with it.

Khodorkovsky went about amassing fortunes as many oligarchs have. He started out importing computers in the Gorbachev years, and then moved into banking. He acquired the assets that became Yukos first through an investment tender in 1995, and then gained control as part of the 'loans for shares' sale in 1996.[12] So far, so typical of the oligarchs.

But once he had got the oil and gas assets he set about building a serious company. He may have acquired his assets in a very Russian way, but he then decided to turn them into something of substantive and lasting value, and to build new pipelines to China as an independent company. His problem was a different one: he was not willing to play by Putin's rules. He wanted to determine his own strategy, including pipelines to China, and he developed dangerous political ambitions to challenge Putin head on. Putin had no concept of a company independent of the state, and especially not in the oil and gas sector.

To do all this and to secure Yukos's future, Khodorkovsky set about finding a foreign (US) buyer for all or part of its businesses, or at least a US buyer for Yukos's oil, breaking the export monopoly. This threatened not only to keep assets out of Putin's control, but also to bring Western corporate governance and accounting to bear on Russian oil and gas. A core element of exploiting the resources for the benefit of Putin and his associates – secrecy – would be partially lost, and as a former KGB and then Federal Security Service (FSB) man, Putin knew all about the value of inside information. That the 'rule of law' was applied – in this case, tax law – merely served to underline an important truth in Russia: that property rights were defined by their contingent utility to the state and Putin, and not by independent courts and judges. The law was Putin's to apply.[13]

This flexible use of the law as a policy instrument was extended to other parts of the oil and gas industry as Putin reasserted control. Shell found itself having to cede parts of its Sakhalin-2 project as 'environmental' law was applied. Notwithstanding the risible idea that environmental protection was taken seriously in Putin's Russia, or indeed by any of his predecessors, it showed how flexible means could be applied by an iron determination to achieve an end of gaining control.

The BP experience illustrated how this could play out as a mix of achieving these ends and the opportunism in the process of achieving them. John Browne, BP's CEO at the time, had recognized that the company needed new reserves and a new frontier.[14] The company had been built originally on the basis of Iranian oil. It had then turned to Alaska, before moving on to the North Sea. As these reserves were depleted it badly needed to replace them. After the collapse of the Soviet Union, and in need of revenues as the oil price continued to fall in the 1990s, Russia welcomed him and BP's expertise.

Browne had few illusions about what he was getting BP into, and BP was quick to recoup its investments, taking significant dividends out in the 1990s.[15] With Putin's arrival, its position began to gradually deteriorate. As a master of opportunism in the pursuit of his objectives, Putin played BP brilliantly. It started with the gas fields at Kovykta in Eastern Siberia, in which BP held a 50% stake through its oil and gas joint venture, TNK-BP. These required a route to market (which might eventually have been China), but Putin had granted Gazprom a pipeline monopoly. Without access to a Gazprom pipeline, BP could not exploit the reserves, and if it failed to produce, it violated its licence. A brutal power play, plus the convenient use of contract law, did for BP. Gazprom acquired the gas field for virtually nothing.

BP played Putin's game to try to hang on to what it had. It provided a very convenient cover for Putin's takeover of Yukos's assets. Rather than simply expropriate them, Putin again used the cloak of legality, Russian style. The Yukos assets were seized to pay for the tax deemed due by the courts. To appear above board, the assets were auctioned. An auction in Russia does not, however, necessarily mean an open, competitive process. Every potential bidder knew the score. Putin had by this stage lighted upon Rosneft as his chosen oil industry vehicle, and Rosneft was effectively designated the winner before the auction took place. BP was the only other bidder, lending a convenient veneer of respectability to the process.

BP had one more part to play in Rosneft's acquisition of the TNK-BP oil assets. The process started with a falling-out with the partners in TNK-BP. The oligarchs in TNK knew the rules of Putin's game and they played them well. In the end, after lots of twists and turns, including the then head of BP Russia Bob Dudley losing his passport and fleeing from Russia, Rosneft acquired BP's assets in exchange for 20% of Rosneft shares. BP now found itself a minority shareholder in a key Putin company.

Getting control meant not only ownership but also controlling the boards of the key companies, Gazprom and Rosneft, and this meant appointing Putin's loyal lieutenants. These came overwhelmingly from two overlapping groups: his former KGB colleagues, and those who had worked with him in his St Petersburg years. A remarkably small group was to stay the course with him through the next decade and a half, filling the central political supporting roles and the top energy positions. Others acquired riches and offshore bank balances that made them billionaires.[16]

Putin started with Gazprom, replacing the long-time chairman with his St Petersburg associates, Alexey Miller and Alexander Medvedev. Within a short period of time many of Gazprom's board were actual or former FSB people. The interlinking set of relationships, and especially mutual dependencies among the new ruling class, the *siloviki*, made the formal roles less important.[17] Putin's regime was more of a clan than a functioning institutional structure, where personal loyalty counted most. As it acquired control of more assets and the state's revenues, it displayed the character of what Karen Dawisha describes as a *kleptocracy*.[18]

None of this takeover of the energy industries followed a rigid or formal plan. Without the checks and balances of a functioning democracy and an independent judiciary, Putin always had the scope for opportunism. Sometimes international oil companies would be welcomed, sometimes they were tempted as partners, and sometimes they were rewarded. They could never be sure, and this uncertainty increased Putin's grip. Companies seeking Russian contracts would obviously want to please the Kremlin, and company executives would be encouraged to argue Putin's case – for example, in the case of South Stream, the Russian pipeline bypassing Ukraine to the south into Bulgaria.

The lure of Russian oil and gas was just too tempting for the main European companies, and to be fair, Russia was just another resource-cursed and corrupt authoritarian petrostate – one among many they had to deal with. In Russia's case it was all about getting back control and then the exploitation of the financial benefits by a small elite. In most Middle Eastern countries, control had already been regained back in the 1970s, as we saw in Chapter 5.

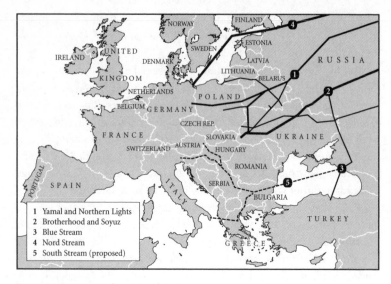

Figure 6.2 Russian pipeline network to Europe

Source: Aurora Energy Research

Playing Putin's cards: Ukraine and its consequences

In Putin's Russia, the gap between politics and economics never really existed, and to the extent that it did, it was gradually closed. But whilst the money mattered – it was the currency for bribery, corruption and ultimately conformity – Putin's strategy was never that of the robber baron. As his thesis had made clear way back in his St Petersburg days, money and control were means to an end: the reassertion of Russia's role in the world as he saw it, and the respect he thought it deserved. Putin's KGB background instilled in him values and a historical perspective that had at its heart the notion not only that Russia was a great nation, but also that the collapse of the Soviet Union had been a disaster that ought to be rectified.

The gradual expansion of the EU, and especially NATO, to Russia's borders after the Soviet Union collapsed is not something Putin ever accepted, and his strategy has always been revisionist. As the 2000s progressed, this 'historical wrong' done to Russia showed every sign of getting worse. The Rose Revolution in Georgia in late 2003, and then the Orange Revolution in Ukraine in late 2004, and the prospect that both countries might join NATO and, in Ukraine's case, the EU as well, horrified Putin. It was not just that these countries were well within what he thought of as Russia's sphere of influence and might slip their anchors, but that the eventual horror might be an Orange Revolution in Russia itself. Indeed, after the election in 2011 this looked a distinct possibility.

Putin started with brute force against Georgia. He could afford to do so, with the oil price now having tripled since he came to power. He annexed Abkhazia and South Ossetia, the result of military action by the Georgians that Russia had cleverly provoked. Russian troops got very near critical pipelines from the Caspian Sea to the Turkish coast, and this too helped to put pressure on other Caspian states that might have been tempted to sign up to the EU-backed proposed Nabucco pipeline and a further outlet to the West, and hence escape from the clutches of Russia.

The intimate relationship between energy and politics played out more directly in Ukraine.[19] Like many Russians, Putin had never accepted that Ukraine was a sovereign state. For many it was part of Russia itself. The Rus people, descended from the Vikings, had Kiev as their capital back in Russia's early history, and it took on an almost mythical part of the narrative of the Russian peoples.[20] It was Ukraine that eventually destroyed the remnants of the Soviet Union when it voted overwhelmingly for independence in 1991, and failed to support Gorbachev in his final efforts to keep the Soviet Union alive as a diminished association of Belarus, Ukraine and Kazakhstan.[21]

Independence merely in name could, however, keep the bond with Russia alive. What changed the relationship was the Ukrainian Orange Revolution in 2004. Despite a botched attempt to murder the Orange leader, Viktor Yushchenko, which left him facially disfigured, the uprising succeeded. Yet it was soon mired in the politics of gas – both internally and with Gazprom. Internally, the Ukrainian elite battled for control of the pipelines and contracts; externally, Russia threatened to cut off the gas supplies.

It was the cutting-off of supplies that alarmed the Europeans, who had been slow to come to the aid of the Ukrainians. In 2006 the first wake-up call was received, to be repeated again in 2009. Russia clawed back an element of influence over Ukraine's politics, and got their man Viktor Yanukovych elected president again, although the gas controversies did not go away. What complicated matters further was the naval base at Sevastopol in Crimea, crucial for the Black Sea Fleet and Russia's route to the south-east and the Balkans.

Crimea was annexed by Catherine the Great in 1783, and had been a long-running source of conflict. There was even a Crimean War in the nineteenth century to limit Russian designs on the Ottoman Empire. Stalin had deported the Crimean Tartars in the Second World War, and Russians dominated its economy. When revolution broke out again in Kiev in 2014, following the tussle between Putin and the EU over which trade block Ukraine would join, Putin resorted to force, annexing Crimea. 'Little green men' appeared to orchestrate the 'referendum' in Crimea – in effect, 'deniable' Russian soldiers and special forces. There then followed a Russian-supported rebellion in the Donbas in south-east Ukraine, leading many to expect that a gradual destabilization and annexation of parts of Ukraine would follow. A Russian BUK missile was widely believed to have shot down a Malaysian plane in 2014, and full-scale war looked ominously likely. All this could be afforded, and particularly the cost of supporting the Crimean economy now cut off from its Ukrainian hinterland, because the oil price was above $100 a barrel.

The Europeans responded more forcibly than Putin had anticipated, notably German Chancellor Angela Merkel. The 'special relationship' that had existed through the twists and turns of the Cold War, *Ostpolitik* and reunification did not stand the immediate test of Ukraine, and the Europeans not only saw off the South Stream pipeline but also imposed sanctions. The Europeans were not to know that the sanctions would come into play at a very vulnerable time for Russia when the oil price collapsed. At $100 per barrel, Russia could take the sanctions hit. At below $50 it was a different story.

The pivot to China

Putin was now boxed in. Sanctions from the US and Europe were taking their inevitable toll, and the Europeans were turning their back on unreliable and politically tainted Russian gas. LNG terminals were sprouting up across Europe; US shale gas was displacing exports to the US from Qatar and elsewhere, and hence these became available to Europe. It looked increasingly like the bear hug of Russian gas on Europe was not going to deliver the long-term revision of the borders and spheres of influence that Putin wanted.

Putin's reaction was twofold. In addition to greater aggressiveness towards the EU, including financing far-right parties in weaker European countries, and the usual tricks reminiscent of the KGB, Putin turned to the other great authoritarian state, China. The Russian–Chinese relationship has a long and difficult history. After the end of the Chinese civil war in 1948, the two Communist states coexisted, with Mao sheltering under Stalin's umbrella. But the relationship soured under Nikita Khrushchev, especially in the context of China's requests for nuclear technology transfer, and at a low point the two countries fought over the border. As China's rapid economic transformation gathered speed after 1980, the Soviet Union was stagnating and then imploding. Indeed Russia's collapse proved a template for all that China wanted to avoid, and at Tiananmen Square in 1989 it displayed an altogether more robust response to the threats to its Communist leadership, just when the Soviet borders were crumbling in Europe. Mikhail Gorbachev allowed the Berlin Wall to come down; Deng Xiaoping crushed the student protesters with tanks.

China's enormous appetite for raw materials and its burgeoning population contrasted with Russia's need for export markets and the empty and open spaces of Eastern Siberia. Russia's relationship represented a mix of fear and opportunity: fear of being swamped by Chinese immigrants, and the opportunity to sell its natural resources. The annexation of Crimea, and Russia's involvement in the civil war in the Donbas, changed this: now Russia needed China as its European markets for gas turned sour and as sanctions bit. It was no longer a relationship of equals.

There followed detailed negotiations about new pipelines and gas supplies, and in 2014 the two countries signed two 'deals': Power of Siberia, a thirty-year contract for the provision of 38 billion cubic metres (bcm) of gas each year starting from 2018 from Western Siberia through a new pipeline; and separately, a framework agreement (not a contract) for 30 bcm of gas a year through the Altai pipeline. As is to be expected, the terms remain opaque and it is likely that China will continue to drive a hard bargain. Putin presented the outcome as a triumph, and as a poke in the eye to Europe, but in reality the gas would come from fields not supplying Europe and in any event would take time to materialize, not least because the pipelines need to be built.

The Russian budget and the long-run consequences of falling prices and new technologies

By late 2014, Putin faced what for him was a new world. For the first time he was confronted by a sustained fall in prices, just as he was engaged in a substantive and costly military expansion programme. With oil and gas having comprised 70% of the government's revenue, a wholly new situation arose: a lot less money, sanctions and a serious economic recession. Lower volumes compounded the lower prices: Europe had been buying less Russian gas for some time. This time there was no industrial base to fall back on. The years of plenty had been squandered. There was comparatively little to show for them, and Russia was exposed.

In the short term, the limited reserves built up in the good years could tide Putin over. Add in a lot of nationalist rhetoric and a defiance of the world outside, and his popularity held up. But as it began to emerge that the prices were not going to go back up soon, and as the financial reserves depleted, hard choices emerged – not least the financing of Crimea, the military budget expansion, and then his Syrian adventures. Putin found himself in the same boat as Gorbachev and Yeltsin: a resource-dependent economy is dependent on the resource prices.

The option of diversification had not been taken in the good times, and it is probably too late now. It is hard to think of any Russian products that would stand the test of international competition except military ones. Indeed it is hard to name any other non-military Russian products at all. As far as the new technologies discussed in Chapter 3 are concerned, Russia is nowhere: nowhere in robots, 3D printing, solar, or even mainstream software and data.[22] The resource curse has gone deep: Russia has serious social problems, great inequality and poor life expectancy. Much of its population is dependent on state benefits, and the good years of ever-more revenue to pay off the population are now effectively over.

It is not going to get better anytime soon. Putin has no answer to the carbon challenge, except to ignore it; no answer to the coming of low-carbon technologies; and no companies able to compete in these new markets, except perhaps in nuclear power. International pressure to 'do something' about emissions will mount, and the decarbonization agenda will, as described in Chapter 2, bear down on demand for Russian energy exports. Russia is stuck with its reliance on its fossil fuels, and therefore on the future markets for these fuels.

The pressures are already reflected in the European gas markets, on which Russia depends. The combination of more and more LNG, the growth of renewables, and the aftermath of the financial crisis in the last decade have dragged demand down. Gazprom now has to contemplate the Saudi oil strategy: the prospect of having to engage in a serious price war to see off yet more LNG. Indeed, price reductions have already been forced upon Gazprom. The hard-won reputation of reliability of supplies even during the worst moments in the Cold War was broken in the two major Ukrainian interruptions. Credibility will be almost impossible to restore for years to come. For Poland, and the Baltic states in particular, the threat of another Russian bear hug drives them to nuclear, coal, LNG and even renewables.

Further out, and in a possible post-Putin world in the middle of the next decade, there are the opportunities presented by the Arctic. Russia has laid claim to the biggest share of its resources on the Arctic sea floor; the waters are shallow and the ice is retreating. Developing these resources will demand new technologies, currently beyond Rosneft and the other Russian state-owned companies. Its pariah status in Europe and the US will further hamper the development of this option, and the more it gets left behind, the greater the risk that climate policies and low-carbon technologies will undercut its very high costs. Arctic resources are marginal ones, dependent on high oil and gas prices, higher than the medium and longer term is likely to deliver.

Two Russian futures

The great paradox for Russia is that it has an amazing abundance of natural capital – just when the value of the energy-related elements may be slipping away. Nature has dealt a wonderful hand, but the Russians have in the twentieth century, and especially in the early twenty-first century, squandered much of it. And it has very little to show for it. Energy has kept Putin's regime afloat, but at a price of both increasing the reliance on oil and gas and leaving the underlying position worse, as the falls in prices after late 2014 have revealed. Although it has the Arctic, shale deposits and lots and lots of existing resources, even if it invests it may be for a product the world no longer needs in such great quantities.

What then happens? There are several scenarios. The first, and perhaps most likely, is that Russia gradually collapses back into the sort of economic chaos seen in the 1990s. It might just be sclerotic like the late Brezhnev years, but it could be nasty. Putin and his associates have nowhere obvious to hide. Autocrats tend to die in office for fear of the consequences of losing power and others coming after them. There are many who would like to tackle Putin and his record.

With less money, and with what is left of the economy after almost two decades of Putin's rule, hanging on to power may be altogether more difficult. External 'enemies', nationalism and patriotism are what saved Stalin in the 1930s after the damage wrought on the economy by his brutal dictatorship, which was so terribly exposed when Hitler invaded. Putin has lots of 'enemies' to focus on, and a sense of national grievance and humiliation to ground his nationalism – just as China has colonial humiliations, Taiwan and the South China Sea. For Putin, the possibilities are endless: the rest of Ukraine, Estonia, Latvia and even Lithuania, neutralizing Poland's Western orientation, Chechnya, Kazakhstan, Georgia and Azerbaijan are all areas which were at one time or another under the control of Russia and then the Soviet Union.[23]

If the Russian economy sinks back slowly, there will be a tension between the need for external enemies to play out the Russian nationalism theme and the lack of money to pay for further adventures. The optimistic story is of a country which turns away from Putin's statism, reforms its markets and opens itself up to the rule of law, courts, freedom of the press and so on, and in so doing becomes one of the great economic opportunities left in the world, after China's transition slows and then ends.

Russia has an enormous amount going for it. In a world with a population that will possibly go beyond 10 billion people, and one that is warming partly as a result of burning all those fossil fuels, the great Siberian expanse is teeming with opportunities. Russia still has a technical educational system of some standing, and almost all of the country is empty. Add in Arctic navigation as the ice melts and it is hard to think of a better-placed geographical area on the planet. Technology plus resources plus space could be a winning combination. Yet the gap between the possible and the likely is a big one. The most likely predictable surprise is that Russia slips back towards a global mid-ranking country, and its economy stagnates.

CHAPTER 7

China
THE END OF THE TRANSITION

If you are looking for one of the main causes of the great commodity super-cycle, the place to start is China. China's economic transformation has been dramatic, but not miraculous. Unlike miracles, it has a rational explanation and, again unlike miracles, it has an end as well as a beginning. Other countries have had transformations in the past, and no doubt there will be others in the future, probably including India.

What makes China different is scale. Everything about China is 'big'. China has more than a billion people; it has been doubling its economy every seven to ten years since the process got going after the death of Chairman Mao. It dominates the world coal market and has had a major impact on oil and a host of other commodities. It has caused massive domestic pollution and massive carbon emissions. It is an economic triumph, and in many respects an environmental disaster. It is at the centre of two of our three predictable surprises – the commodity super-cycle and the carbon constraints. It may or may not be a core player in the new technologies.

To understand how energy markets may shape up over the coming decades, and to understand whether climate change can be limited, China matters in a way that many other players do not. A China continuing to grow at 7% per annum is a very different proposition to a China that follows the Japanese path after its transition ended. Fortunately for the climate and the environment, an economic slowdown and corresponding falls in its energy demand growth are likely to be the outcome. Indeed, both may have already begun.

The first task is to understand the dynamics of China's transformation. Why did it result in the commodity super-cycle? How will it respond to the challenges of the economic slowdown? What will be the impact of its search for overseas supplies of natural resources – from the Middle East, Australia, Africa and now Russia? As a corollary of this overseas expansion, how will the militarization of the South China Sea, the push out into the Indian Ocean, and China's increasing global power projection play out? Together these set the scene for the outlook for China's future demand growth, its impact on energy markets and its carbon emissions, and its ability to cope with the end of the commodity super-cycle, decarbonization and new technologies.

The historical legacy

Economic growth does not happen spontaneously. It has a historical context, a global context and a domestic set of causes. China is no exception. It has at various times in its history given rise to several of the world's great empires, with a distinct culture, art and political organization. The Ming and the Qing periods stand alongside the other great civilizations in history that have had their time in the sun – the Ancient Greeks, the Romans, the Incas and the Mongols. But, unlike these other examples, China has endured, with lots of ups and downs along the way. It has withstood Mongolian invasions, repeated Japanese incursions, struggles with the Vietnamese and British colonialism.

This long history gives China a sense of greatness, and it also conditions its transformation. The humiliation in the nineteenth century, with the burning of the Old Summer Palace in 1860 and the Opium Wars with the British (with American assistance on occasion), and the brutal rape of Nanjing in 1937 and the horrors of the subsequent invasion by the Japanese during the Second World War,[1] are among the most vivid and painful events in China's more recent history. There are other historical ingredients and these have been combined to create a 'National Humiliation History', with its nationalist consequences now witnessed in disputes with Japan and its actions in the South China Sea.[2]

After the overthrow of the last emperor in 1911, the republic was led by Sun Yat-sen and then by Chiang Kai-shek's nationalist government from 1927. He was challenged by Mao, with the Communist insurgence eventually triumphing in 1948, forcing Chiang to seek refuge in Taiwan and in the process creating a political sore yet to be healed in the tensions between China and Taiwan over its sovereignty – one that remains core to Chinese foreign policy today.

Mao's regime created the conditions for the eventual transformation in a largely negative way. Mao showed the limits of dictatorship, Communist style. From the 1950s to the end of the 1970s, his 'Great Leap Forward' and 'Cultural Revolution' resulted in perhaps as many as 70 million deaths, making Mao one of the bloodiest of all the twentieth-century dictators, and ranking him in the super-league of brutality with Hitler, Stalin and Pol Pot.

What these terrible political initiatives demonstrated was the utter failure of complete central planning and Mao's attempt to change the nature of society and with it human nature. In both cases the parallels with Russia and the Soviet Union are obvious. Mao took *Gosplan* to a wholly new level in a predominantly agricultural setting, beyond even Stalin's assaults on the richer peasants, the kulaks, and the resulting famine in Ukraine in the 1930s. Mao's Cultural Revolution represented an attempt to eliminate the educated middle classes, and make proletarians (or in China's case, peasants) of everyone. His cult of personality, and the ruthless persecution and murder of other leaders, again eerily echoes Stalin.

It was out of this terrible experience that the modern China has emerged.[3] Although the post-Mao era is seen as a break with the past, it is not a clean one: the dictatorship of the Communist Party remains; Mao is still revered; opposition is still crushed (as at Tiananmen Square in 1989, and repeatedly in Tibet and the Uyghur region in the north-west); rival leaders are still purged, jailed and sometimes executed; and individual freedom is limited through mass censorship and control of the media. The difference between the last three decades and what went before is the economic success, but it is an illusion to think that the Chinese model has given way to capitalism, and that with capitalism and economic growth, liberal democracy will follow. Chinese state capitalism is not the sort of market capitalism familiar in the West, and neither China nor Russia should be assumed to be transitioning to the US or European democratic models – a common mistake made by Western liberals and political leaders. The 'end of history' is not necessarily a Western liberal democracy.[4] This applies particularly to its energy sectors, from nuclear and renewables to oil, gas and coal. State capitalism is not the same as market capitalism.

It is relevant that the key architect of the transformation was one of Mao's close associates, Deng Xiaoping, and that the ensuing leaders have been drawn largely from the children of Mao's Communist regime, the so-called princelings. None has escaped their past, and all have jealously protected the Communist Party's grip on power. They have, from Deng onwards, taken the view that material success and nationalism are the route to maintaining the loyalty of the population and providing the basis for China's reappearance on the world stage, and for the righting of the historical humiliations inflicted by the British, Americans and Japanese. Chinese leaders look to the collapse of the Soviet Union as a case study in what can go wrong (and in many respects Putin holds a similar view of the Soviet Union). Coincidental with the fall of the Berlin Wall, China had Tiananmen Square. China's leaders have little appreciation of *perestroika* and *glasnost*. Where Gorbachev caved in to the demonstrators, the Chinese crushed them.

The transformation wrought by Deng was more gradual than hindsight suggests, and there was no economic transformation blueprint. It began with free-enterprise zones and the attempt to build up small-scale experiments (again, as it did in Russia in the 1980s, notably in St Petersburg). Hong Kong had sat there as an example of what could be achieved through openness to trade. China did not *decide* to become the world's greatest energy-intensive exporter: it happened gradually, in an opportunistic way, and because a number of core necessary conditions were in place.

These conditions were internal and external. Internally, China had an almost limitless amount of cheap labour in the countryside. Mao had seriously held back its urbanization, and there was a natural rebalancing towards the cities after his excesses. China had lots of coal too. What it lacked was internal demand, and therefore the only route to economic growth lay through exports and trade. It was a pattern familiar in post-Second World War Germany and slightly later in Japan, South Korea and Taiwan. Transitions are almost always export-led.

China's economic model

An export-led growth strategy requires others to have a corresponding demand for imports. The Chinese transformation just happened to coincide with the great late-twentieth-century boom in the US and Europe. After the dismal 1970s, from 1980 the Reagan and Thatcher era kicked off a revival of consumption in the West, and it was further underpinned by the communications revolution (again a key ingredient in explaining China's growth). As the US embraced a new optimism, consumption could be sustained at ever-greater levels on the assumption that the future would be bright, and hence higher future incomes could pay off higher current debt levels. This was as much a part of the commodity super-cycle as China's satisfying of it.

The great consumer-led boom in the West was accompanied by the abolition of exchange controls and many of the constraints on borrowing, particularly for mortgages. Reagan went for a very Keynesian demand stimulus, backed up by military expenditure. A boom ensued, leading to the first of a series of stock market crashes in 1987. The result was a lurch towards ever-looser monetary policies, which were to last for three decades. Notwithstanding the Japanese crash in 1989, US and European stock markets powered on until 2000 when they too crashed, never to regain their real value again through to 2016. Yet more stimuli were applied – negative real interest rates, George W. Bush's tax cuts in the US, and Gordon Brown's splurge of public expenditure in Britain, and similar if more constrained moves across much of Europe. After 2000, consumption boomed again, this time with an asset bubble on a new and frightening scale. This started to burst from 2006, leading to the credit crunch and the sub-prime lending crisis, and to even more stimuli and even lower interest rates, and eventually to outright printing of money (quantitative easing, QE). Nominal interest rates even went negative.

Without this demand from the US and Europe (together nearly half of world GDP), the Chinese transformation would have been an altogether more muted affair. China's growth was focused on its eastern coast, and on large-scale energy-intensive manufacturing in steel, fertilizers, cement, clothes, electrical goods, and so on.

To meet this demand, China put its cheap labour together with abundant coal. Between 1980 and 2012, the 'Great Migration' led to more than 200 million people flocking from the countryside to the cities, and China's urban population tripled.[5] Two numbers stand out from the perspective of coal and steel, the commodity super-cycle and energy. Coal-burn went up from 610 million tonnes to more than 3,500 million tonnes over the same period; by 2000 China was building more coal-fired power stations each year than the UK's total installed capacity. And China's production of steel rose from 37 million tonnes in 1980 to 823 million tonnes by 2014.[6]

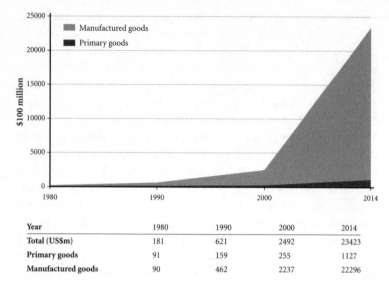

Year	1980	1990	2000	2014
Total (US$m)	181	621	2492	23423
Primary goods	91	159	255	1127
Manufactured goods	90	462	2237	22296

Figure 7.1 Chinese exports value by category of goods (US$100 million)

Source: National Bureau of Statistics of China, *China Statistical Yearbook 2015*, http://www.stats.gov.cn/tjsj/ndsj/2015/indexeh.htm

Transforming world energy demand

What this transformation meant for energy markets is extraordinary. Until 1995 China had been an exporter of coal; yet by 2015 it was responsible for half of all world coal trade.[7] Its demand for oil rose dramatically too, from less than 100 million tonnes in 1980 to almost 500 million tonnes in 2012 – that is, less than 2 mbd in 1980 to more than 9 mbd in 2012.

For the first two decades of its economic transformation China was fortunate that world commodity markets were benign: oil fell from $39 in 1979 to less than $10 in 1999, without taking account of inflation. In other words, the first two decades of China's transformation were ones in which it could combine its *cheap* labour with *cheap* energy and *cheap* commodities.

By 2000, China had arrived on the world stage, and the process of offshoring manufacturing from the US and Europe to China was well underway. The US and Europe could not compete with China's labour costs, which were a fraction of those back home. China faced few large-scale strikes, a continuous inflow of labour, and none of the onerous health

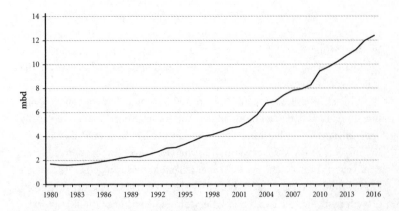

Figure 7.2 China's oil consumption, 1980–2016

Source: BP, 'Statistical Review of World Energy 2016', June 2017, http://www.bp.com/statisticalreview

and safety and other labour laws and regulations that drove up costs in the West. It could brush off the many and terrible industrial accidents and the large scale of deaths and health damage caused by its coal industries. Almost everything in China was cheaper, and hence the manufacturing in China of everything from iPhones and laptops to steel and even cement became a competitive necessity for global companies. It was the main act in the great globalization of trade from 1980 to 2008.

Although these benign conditions for China gradually started to unwind in the 2000s, it seemed initially that, if anything, conditions were becoming even more adverse in the West. Rising oil prices hit the US and Europe just as badly. Furthermore, as the economic crisis began to unfold in 2007–08, China's authoritarian state did not have to worry about the niceties of legal liabilities and the mobilization of the banks and other assets. It could go for financial and monetary stimuli on a massive scale. And it did, notably in infrastructure and property.

It was not until the late 2000s that the first serious cracks began to appear. China's massive infrastructure spend to head off the economic crisis bought time, but at the expense of a correspondingly massive misallocation of capital. Its property market cracked, and pollution began to cause unrest and threaten loyalty to the Communist regime. The supply of cheap labour was slowing, and the one-child-family restrictions were feeding through to an ageing population. Its neighbouring Southeast Asia rivals had cheaper labour, whilst the lumbering giant of India had these and other advantages, including democracy. The debts were piling up, notably in the opaque accounts of the state-owned enterprises.

Whilst China's leaders could buy time with stimuli, they could not change one fundamental: the falling-off of demand for its exports in the West as the economic crisis cut hard into their economies. Neither could they change the impacts of the shale revolution in the US. Lower import demand, and relatively and absolutely much cheaper US energy costs, together threatened China's model. Its solution has been to try to move from an export-orientated growth model to one based on higher domestic consumption. It is a formidable challenge, and there are few historical examples of such a successful transformation. Japan has failed, slumping to zero growth after 1989, and staying in the very-low-growth zone in the quarter-century since. Germany may be a more encouraging example, but it has democracy and has been content with 2–3% growth in recent decades. Further back, Britain's transformation from the leading global trading nation to its more modest position today suggests that the switch to domestic demand may not be easy. Moreover, this transition was punctuated by the devastating world wars.

The slowdown in Chinese growth has now begun to happen. The end of the commodity super-cycle described in Chapter 1 is in part the result. China may claim that it is growing still at 6–7% per annum, but the numbers at best disguise more fundamental problems. Its looming debt crisis threatens a more sudden and sharp contraction. Some of its state-owned enterprises have serious financial problems too. Looking ahead there are good reasons to assume that this structural break in China's commodity demands may be permanent. Transitions are transitions – they start and they end.

China's global suppliers

The enormous scale of the energy demands necessitated an equally enormous development of resources. Some of these have been national. China produces lots of coal and a lot of its oil requirements. But in the 2000s, it needed new supplies, and it needed to make sure that these would be secure.

To achieve this resource security, China looked towards an overseas development model, which took it into direct investments in energy (and also agricultural) projects. Africa was a particular destination. China was an attractive investor. It was not concerned with human rights or the behaviours of dictators. It had no agenda to promote democracy. It had money and technical expertise, and was in for the long term. From the perspective of a number of African states, what was not to like about this, compared with the scruples of Western democracies and the companies they regulated? As illustrated in Figure 7.3, countries like Nigeria, Angola, the Democratic Republic of Congo and South Sudan received lots of Chinese interest, and elsewhere minerals were sourced from Zambia and the Democratic Republic of Congo, together with broader investments in Kenya and Uganda.

China looked to the Middle East for resources too. Again, its lack of concern about the regimes it dealt with gave it opportunities others could not so easily exploit. Iran, for example, supplied nearly 10% of China's oil imports in 2014, in the context of US- and UN-backed sanctions. As Figure 7.4 shows, it had deliberately diversified its sources of supply. In its own backyard, China had Indonesian coal and a major supplier in the shape of Australia (respectively 34% and 38% of its steam coal imports).

As gas became more important, both to cut pollution from its power stations and to try to match the revolutionary US shale gas impacts on its competitiveness, China turned to Russia for pipeline gas. There had been an energy relationship for many years, with 11% of Chinese oil imports coming from Russia in 2014.[8] Yet the relationship had not always been easy. It had started off well at the end of the 1940s when Mao came to power, but it had soured after Stalin's death in 1953 and on occasion came close to a major war.

What changed the Russian–Chinese relationship was the souring of Putin's relationship with Europe and the US. Europe has been the preferred market for Russian gas for several reasons. As we saw in Chapter 6, the gas fields were developed to the west of the Urals with Europe in mind. Until recently, Europe had a steadily rising demand, it paid on time, and it gave Russia influence where it has felt it needed it, and leverage over its former Soviet backyard. Russia has through most of its history seen itself as a European power. But after the annexation of Crimea, the Russian-backed civil war in the Donbas and the consequent sanctions, Russia has turned to China as an alternative market. It has done this from a position of weakness: Russia needs the revenues from China more than China needs Russian gas.

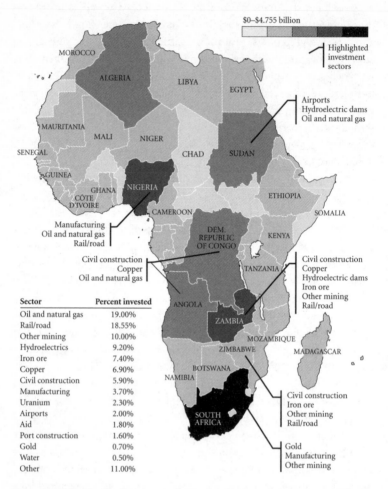

$0–$4.755 billion

Highlighted investment sectors

MOROCCO

ALGERIA

LIBYA

EGYPT

Airports
Hydroelectric dams
Oil and natural gas

MAURITANIA

MALI

NIGER

SENEGAL

CHAD

SUDAN

GUINEA

GHANA

NIGERIA

ETHIOPIA

CÔTE
D'IVOIRE

CAMEROON

SOMALIA

Manufacturing
Oil and natural gas
Rail/road

DEM.
REPUBLIC
OF CONGO

KENYA

Civil construction
Copper
Oil and natural gas

TANZANIA

Civil construction
Copper
Hydroelectric dams
Iron ore
Other mining
Rail/road

ANGOLA

ZAMBIA

MOZAMBIQUE

ZIMBABWE

MADAGASCAR

BOTSWANA

NAMIBIA

SOUTH
AFRICA

Civil construction
Iron ore
Other mining
Rail/road

Gold
Manufacturing
Other mining

Sector	Percent invested
Oil and natural gas	19.00%
Rail/road	18.55%
Other mining	10.00%
Hydroelectrics	9.20%
Iron ore	7.40%
Copper	6.90%
Civil construction	5.90%
Manufacturing	3.70%
Uranium	2.30%
Airports	2.00%
Aid	1.80%
Port construction	1.60%
Gold	0.70%
Water	0.50%
Other	11.00%

Figure 7.3 China's foreign direct investment: stock and investment composition in Africa as of 2012

Sources: UNCTAD, 'Bilateral FDI Statistics 2014', http://unctad.org/en/Pages/DIAE/FDI%20Statistics/FDI-Statistics-Bilateral.aspx; Stratfor, 'Chinese Investment Offers in Africa', August 2012, https://www.stratfor.com/image/chinese-investment-offers-africa

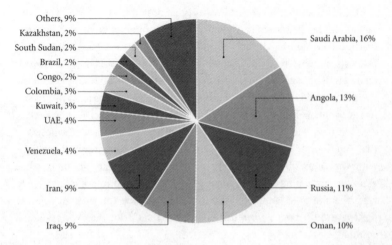

Figure 7.4 China's oil imports by source, 2014

Source: EIA, China country profile, http://www.eia.gov/beta/international/analysis.cfm?iso=CHN

Dependency works both ways: OPEC, Russia and other producers may have held the whip hand as prices strengthened in the 2000s, but falling prices and revenues reversed this. Add in new pipelines along the new Silk Road – the One Belt–One Road strategy – to Turkmenistan and eventually perhaps even Iran, and the great gas reserves developed in Australia to feed the Chinese LNG market, and the imbalance of bargaining power is further shifted in China's favour.[9]

China's blue-water fleet

The One Belt–One Road has a naval dimension too – a twenty-first-century maritime road from the South China Sea through the Indian Ocean and beyond. As China has become increasingly dependent on imported energy, it finds itself in a similar position to the US in the second half of the twentieth century, which had developed a deep vested interest in making sure these supplies keep coming. To date, China, along with the Europeans, has simply relied on the US to keep the Straits of Hormuz and Malacca and other key shipping lanes open. The US needs security too, and China and others could simply free-ride.

This is changing for several reasons, all rooted in the 2000s. The military problems in Iraq and Afghanistan, the economic crisis and its aftermath, and the coming of shale oil and gas all change the US's vital economic interests, as described in Chapter 4. As North America moves towards energy independence it no longer needs the Middle Eastern suppliers in the ways it once did. It can get by without Saudi Arabia if it has to. Whilst it still has geopolitical interests, they are not as economically vital as they were.

This new American detachment has been noticed, and the result is that Saudi Arabia and other Gulf producers have sought better relations with China to fill the emerging potential security gap. To meet these expectations, China needs to move gradually to a position of being able to keep the shipping lanes open by itself and provide the sorts of military hardware Gulf states have relied upon from the US for their own regional conflicts.

As the US retreats, China is developing its own Mahan strategy (that naval power is the route to empire[10]) in its own backyard, the South China Sea, and is seriously thinking through its ability to defend the Strait of Malacca and build a blue-water capacity in the Indian Ocean. To meet these various needs, China is undergoing a rapid naval expansion and building key alliances along the way. It is developing its underwater nuclear submarine base at Hainan Island, militarizing the rocks and coral reefs of the South China Sea (notably the Paracel and Spratly islands), disputing the Senkaku/Diaoyu islands with Japan, and constructing ports, railways and oil storage facilities at the end of its 2,800km oil and gas pipeline at Ramree Island off the coast of Myanmar. It has committed to developing ports and naval facilities in Pakistan, near the mouth of the Strait of Hormuz.[11]

These developments make China an emerging military player in the global energy markets, and bring with them a de facto role in maintaining security of supply. But whereas the US has treated this role as one largely for the benefit of all (including China and Europe), China's intentions may be less benign, especially towards the US. It aims to drive the US Navy out of the South China Sea as a first step, and this explains in part why the Obama administration (and notably his first secretary of state, Hillary Clinton) talked of a 'pivot to Asia'. Should relations with the US deteriorate significantly, this will have serious implications for future global energy security, though probably worse for China with its import dependency than for the US.

China's state-owned energy companies

China's expansion into the energy field has been built on state-owned and state-controlled companies. There is no 'free market' here: the field is deliberately tilted in its favour. As Figure 7.5 indicates, Sinopec, the Chinese National Petroleum Corporation and the State Grid Corporation of China are all in the top ten companies in the world by size in 2015. These companies are ultimately instruments of the Chinese Communist Party and, as with all resource companies, the scope for corruption is considerable. The accounts lack credibility, and Xi Jinping's 'crackdown on corruption' is better understood as a battle over power and control. In a country in which property rights depend on the state and the Communist Party, the very definition of corruption is hard to apply on an individual basis. As with Putin's Russia, corruption is embedded in the very political structure, since who gets what depends ultimately on the will of the leader. It is no accident that President Xi Jinping has used his anti-corruption campaign to eliminate his rivals – just as Putin used Russia's 'tax laws' to get rid of Khodorkovsky and other oligarchs who do not toe his line.

As instruments of the state, companies' finances are washed through those of the government. Indeed, this has been exploited to the extent that they may have a debt exposure on a par with the sub-prime crisis in the US in the last decade.

Fortune 500 rank	Company	Country	Industry	2015 revenue (US$ billion)
1	Walmart	US	Retail	$485.7
2	**Sinopec**	**China**	**Petroleum**	**$446.8**
3	Royal Dutch Shell	Netherlands and UK	Petroleum	$431.3
4	**China National Petroleum Corporation**	**China**	**Petroleum**	**$428.6**
5	ExxonMobil	US	Petroleum	$382.6
6	BP	UK	Petroleum	$358.7
7	**State Grid Corporation of China**	**China**	**Power**	**$339.4**
8	Volkswagen	Germany	Automobiles	$268.6
9	Toyota	Japan	Automobiles	$247.7
10	Glencore	Switzerland and UK	Commodities	$221.0

Figure 7.5 World's ten largest companies by revenue in 2014

Source: Fortune, 'Global 500', 2015, http://fortune.com/global500/

Weaknesses ahead: why China may not continue to
drive energy demand

In retrospect, China's great growth burst probably started to unwind at the same time as the global crisis in 2007–08. Its export-led model had been financed by domestic savings, routed via the state rather than the market into investment. This created an investment-driven economy, to match the consumption-driven models in the US and Europe. China's model depended on overseas demand and the deployment of domestic savings to build the industries inside China to meet that demand.

There are three sets of problems with the model.[12] The first is that the various bubbles that the government created in its attempt to head off the crash in 2007–08 are now mostly unwinding. The second is the medium-term challenge of switching from investment to domestic consumption. The third is the challenge in the longer term posed by new digital technologies as they displace cheap labour, upon which China's competitive advantage has been based. Together these factors call into question the common assumption that China will continue to drive global energy demand, and indeed carbon emissions growth too. They are major reasons to expect the price of fossil fuels, and oil in particular, to be more subdued.

In the immediate term, the Chinese growth rate is officially around 6–7%, and if this were indeed the case, demand for oil, gas and coal should be rising strongly. But the trouble with China is the 'official' bit. To paraphrase the famous quote, there are lies, damned lies and Chinese statistics. Why would anyone believe that the numbers presented by the Chinese government reflect the true state of the economy? In particular, given that the Chinese government regards its legitimacy as resting in considerable measure on its ability to deliver high levels of growth, why would it be expected to honestly tell its population that all is not as promised by the leadership in the latest Five Year Plan?

The reality is that GDP numbers in China are 'political'. There is no such thing as a neutral economic number in China. If the economy is falling back sharply, it would be odd to expect the published figures to make this plain. A key feature of open democratic societies is transparent independent statistics. China is neither open nor democratic, and its statistics are not independent.

So how can we know what is really going on? There are two routes forward: to look at other indicators, and to come up with a causal explanation of how transitions end.[13] First, the other indicators. If the economy is slowing, the obvious place to look is in the imports of oil, iron ore and minerals, and their global prices. The countries that export to China would be feeling the effects. The serious impacts on Australia's mineral sector, on Indonesian coal exports and on tanker and shipping movements tell a different story to the official Chinese numbers, as more generally do commodity prices. If China was going full steam ahead at, say, 7%, so too would commodity prices, shipping volumes and the exporters' economies. But they are not. Seven per cent of the world's largest economy added every year, and hence another doubling of China's economic footprint by 2025, represents a lot more oil, gas, coal and iron ore demand. This is 7% of a much bigger number as a result of all the economic growth that has already happened.

The other approach is to provide a causal explanation of why the Chinese transition may be ending, and it starts with the underlying characteristics of the model. Like all transitions, it requires an internal and an external framework. Internally, the suppression of incomes to generate the investment funds runs up against resistance over time. It throws up great inequalities to the benefit of those who can capture the rents from the financial flows and investments, and comes with inevitable corruption. As with Putin's Russia, rising inequalities result, with Communist billionaires alongside poor peasants and low-paid industrial workers. The model is not a stable equilibrium. It requires a continuous flow of cheap labour from the countryside and continuous financial surpluses to recycle through the state-owned companies. An analogy can be drawn with the famous quote by Citigroup's Chuck Prince about the US credit boom and bust: 'As long as the music is playing, you've got to get up and dance.'[14] But when the music stops, the dynamics seize up: without more labour, labour costs rise and competitiveness falls; without the surpluses, the investment cannot be supported, and the state has to resort to debt financing and monetization of that debt.

As the music fades out, the system itself may be revealed as unstable. As with the collapse in Japan, the first bits to go are the asset bubbles. In China's case these are in the property market, the companies and the banks. This explains why in mid-2015 the Chinese authorities tried frantically to prop up the stock market. As this proves difficult to achieve, the impacts feed through into the economy. The state and, implicitly, the Communist Party are seen to be fallible, undermining credibility. The options for the state start to diminish. It can resort again to massive infrastructure investments in another stimulus package, as it did after 2007–08. The trouble here is that it is running out of projects, and they are not necessarily good investments.[15] If demand is lower in the future as a result of slowing, there is the risk that China may end up with lots of stranded infrastructure assets. It can try to adopt a new and different economic model, for example through the widespread deployment of robotics, but as with Russia and Saudi Arabia it has obvious limitations and is far from guaranteed to be successful. Crucially, it takes time.

As Lant Pritchett and Larry Summers put it, transitions end because they always do.[16] A transition is a dynamic catch-up process. It varies from case to case. But it is still a transition. The obvious question is: to what? The Chinese answer is to a more consumption-driven, middle-income economy. Yet this requires consumers to consume, and this in turn depends on their confidence in the future. For the Chinese, as growth slows, it is far from obvious that savings will fall. An ageing population and the absence of a welfare state encourages a precautionary approach, as it does in Japan.

External circumstances provide shocks and triggers. China got lucky: its great expansion coincided with the great booms in the US and Europe. But luck tends to revert to the mean. The very global economic features which helped its export-orientated model disappeared with the economic crisis in 2007–08. Add in the shock of US shale oil and gas and the increase in US competitiveness that this entails, and the next decade looks a lot less lucky for China. Its biggest customer has other options and can afford to be more hostile to Chinese trade practices, as Trump has been demonstrating. The fact that China has accumulated over $1.5 trillion of US government securities is a problem for China, not the US. The US can, and has, monetized some of these already and sets the interest rate paid on them.

To add to the short-term problems of its sharp slowdown, and the medium-term problems of trying to transition to domestic consumption, China has lots of long-term problems. These fall into three categories: its labour force and welfare; the stifling impact of the Communist Party and its authoritarian leadership; and the fact that it has few friends, in the context of its military expansions, as it tries to exert its regional and global power.

The great advantage its labour force provided of a cheap, placid and urbanizing factor of production is gradually being turned on its head. The one-child policy has slowed China's birth rate, whilst its economic growth has provided the foundation for greater life expectancy. As a result its population is getting older, and within a few decades it will be falling.[17] It will look in demographics rather like Japan and Germany, two other transitional examples. But unlike either of these, China's income per head will still be quite low, probably too low to make the transition to a consumption-based economy, and too low to afford the sorts of health and welfare requirements of this increasingly powerful, older-generation lobby. Where once the rural communities looked after their old and sick, China's great economic migration has undermined its traditional community welfare system.[18]

The second long-term challenge is how to compete in the new technological world, where labour is further replaced by capital, and where production is more decentralized with 3D printing, robotics and AI (as described in Chapter 3).[19] Its great labour force will not only be vulnerable to the squeeze these trends imply, but they may also not be able to compete with the sorts of human capital that counts. Ingenuity, inventiveness, openness to challenge, and entrepreneurship are not things the Communist Party is very keen on and, as the economy slows, there is little evidence that the much-touted transition to liberal democracy will happen in the country. China does not even have an established system of property rights; these were abolished in 1948, and have been replaced by a patchwork of leases and rights. Much more likely is the ever-tighter grip of Xi Jinping and his fellow autocrats, even if punctuated by the odd economic necessity, like relaxing the one-child policy which controlled even the most basic individual family behaviour.

Third, and more dangerous, is the consequence of China's military build-up, and a resurgence in Chinese nationalism that the leadership might embrace as the economic miracle fades. China has lots of enemies to conjure up before its own population, and authoritarian regimes usually turn to nationalism when they face internal divisions. 'National Humiliation' (as in Putin's Russia) engenders a desire for retaliation. China's militarization of the South China Sea, the new 'islands' it is reclaiming at great environmental cost, the stand-off with the old enemy, Japan, and the obvious intent to push the US out of the western Pacific and to assert Chinese power in the Indian Ocean have already provoked the inevitable reactions. China's neighbours might be browbeaten into acquiescence, but not as willing partners. From Vietnam to Myanmar, Indonesia, the Philippines and Singapore, few show a desire to become vassals of a new Chinese sphere of influence.

Although there is still some way to go before China has the naval power and missiles capable of sinking US aircraft carriers, and there is evident restraint in pushing Japan and the South China Sea powers like Vietnam too far, the odds of an 'accident' are necessarily increased by this military build-up.

Together, these short-, medium- and long-term issues suggest that it would be foolhardy to simply extrapolate Chinese energy demand into the future at current growth rates and assume that it will be the engine for pushing world oil demand to 100 mbd and beyond. More likely is that it might follow the US and European trends, albeit at a much lower level of income per head. Demand growth will probably continue, but probably not enough to offset the increases in supply identified in Chapter 1.

The implications for China's geopolitical role

In the new energy world of lower oil prices, decarbonization and new technologies, many commentators see China as an obvious winner. It benefits from lower prices, it has invested in solar and nuclear on a large scale, and it has the authoritarian power to force technical progress. This is a mistake.

Look more closely and these advantages are less obvious. The lower prices not only benefit China. They benefit the US and Europe too. US energy is even cheaper, and it is the relative price of energy that matters for competitiveness. China's solar companies are less than they seem: some near or actually bankrupt, with large-scale debts and mass production of current-generation technologies. For all the rhetoric and the misguided optimism of some environmentalists,[20] China remains overwhelmingly dependent on coal, which has to go if climate change is to be addressed. And it has little domestic gas yet to offset this. Chinese shale gas deposits are large, but in the wrong places and without plentiful supplies of water for low-cost production. It will need a lot of LNG, notwithstanding the proposed pipelines in the One Belt–One Road strategy, many of which may never be built. Its nuclear new-build is based around various sorts of reactor technologies, but the chosen ones (like PWRs) are not necessarily the technology of the future and are potentially very expensive over the next half-century when it will have to compete with lower medium-to-longer-term fossil fuel prices and the new technologies like next-generation solar, as described in Chapter 3. Finally, where the new technologies discussed above are concerned, it is not obvious that China will have the advantage it has enjoyed in manufacturing for the last three decades. Robots and 3D printing lend themselves to more localized and advanced manufacturing, and in the 'second' industrial revolution China's labour market will have few identifiable advantages. It might be dashing for robots, but not necessarily the more intelligent ones.[21]

The new energy world therefore challenges the Chinese model. None of its apparent advantages turns out to be quite what they superficially seem. China may not have the challenges that OPEC and other major oil producers face, but it does have lots of old coal and, in comparison with the US and Europe, it has offsetting problems ahead. Just as its economic model runs into very choppy waters, what at first glance seems a stroke of luck with the advent of falling commodity prices may not feed through into China being more generally lucky in the new energy world that is gradually emerging. The US, not China, looks more like a winner.

CHAPTER 8

Europe
NOT AS BAD AS IT SEEMS

Among the more unlikely winners in the new energy world, with both low fossil fuel prices and new technologies, Europe stands out with the US. Europe has a lot to benefit from: its market of over 500 million people is energy-poor. It has little oil, not much gas, and as a result it is highly dependent on its main supplier, Russia – a country that has turned back into the 'difficult neighbour' it was throughout most of the twentieth century. A fall in oil and gas prices is therefore very good news, and the prospect of a new normal level of prices for the medium and longer term bodes well.

Europe has two other things going for it that are relevant to the energy context. It is one big market, and one big democracy, based broadly on open liberal principles. It has a deep culture. Information is open, ideas are not censored, and as a result it has a core space for creativity. It has great universities, great science and a large educated population. Only the US scores at a comparable level on these fronts. Both Europe and the US have significant and growing inequality, but the consequence of Europe's welfare states, especially in health and education, is that it has a wider pool of talent to draw on. It is not restricted so emphatically to a rich elite, as the US is.

For these reasons Europe *ought* to thrive in a post-OPEC world, based on technology, ideas and innovation. And yet quite a lot stands in its way. Its two recent crises – the refugees and the Eurozone – weigh it down, with a legacy of higher unemployment, especially among the young, lower growth, high debts and populism undermining the very idea of Europe and the European Union project, which has been the main feature of European politics since the Second World War. Its fraught politics could even cause the EU to disintegrate – starting with Brexit.

Yet, even here, things are looking up for Europe. The doom and gloom have been overdone. However, Europe's recent forays into energy and climate policy provide little ground for optimism. Europe has made most of the wrong moves – most significantly into high-cost first-generation renewables, but also failing to build enough interconnectors quickly enough, and failing to complete the internal energy market after a quarter of a century of trying.

How did Europe get itself into its energy mess? What are the implications of the new oil and gas world and the new technologies? How may these radically alter its geopolitical relationships with Russia, the Middle East and Turkey? How will it fare in terms of its claim to leadership on climate change?

European energy policy: the last twenty-five years

European history has been driven, and dogged, by its need for access to energy. The Industrial Revolution led by Britain was coal-based, as somewhat later was Germany's. The industrial heartlands of Britain in the north and the Ruhrgebiet in Germany are synonymous with coal. France, by contrast, had only the coal deposits in the north-east. Italy had none. Later, with the coming of oil, none had any significant domestic supplies until the 1980s, when the North Sea fields were developed, marking out Britain and Norway from the rest of Europe. Natural gas came later too, from Russia and the North Sea.

The historical point is that, unlike the US and Russia, Europe has apart from coal always been relatively energy-poor, and therefore energy security has been a significant preoccupation and strategic weakness for most major European powers. It explains why Britain needed Iranian oil, why Germany has always had to engage with Russia, and why France tried the radical option of very large-scale nuclear power. Beyond Europe, only Japan faced the same energy weaknesses.

Although the development of the North Sea eased some of the pressures, notably on Britain and the Netherlands, it did not solve the energy security problem, and Germany remained particularly exposed. This was reflected in the attitudes to the liberalization and, in particular, the

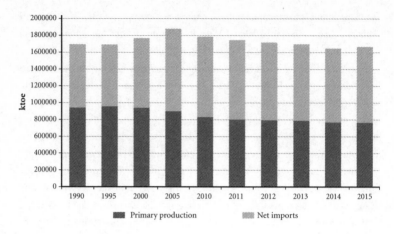

Figure 8.1 Energy balance in EU 28

Note: ktoe, kilotonne of oil equivalent.

Source: Eurostat, 'Energy balance sheets', 2017, http://ec.europa.eu/eurostat/web/energy/data/energy-balances. Eurostat, © European Union, http://ec.europa.eu/eurostat/web/main/home, 1995–2017. Responsibility for the adaptation lies entirely with Aurora Energy Research

Europeanization of energy markets from towards the end of the 1980s. German, French and Italian companies fought a rearguard action as national champions and proxies for their governments. It was reflected too in Germany's *Energiewende*, announced unilaterally after Fukushima, but with its origins in the earlier Chernobyl nuclear disaster.

Grand-scale politics played a part in the energy relationships. Germany's weaknesses led it to turn again and again to Russia, from the Hitler–Stalin pact referred to in Chapter 6, through to *Ostpolitik* and the Nord Stream pipelines. Its 'special relationship' went back a long way. In 1913 Germany was Russia's most important trading partner, and by a very wide margin.[1] Britain, with its energy abundance, could take a different tack: opposing the Soviet Union, and promoting liberalization. France tended towards the German view and continued to engage with the Soviets, whilst Italy needed all the friends it could get for its energy supplies, including both Russia and Libya.[2]

The big players came to the liberalization programme with very different starting points and priorities. Creating a single market in energy was, in general terms, a natural extension of the 1985 White Paper, 'Completing the Internal Market'.[3] This initiative, driven by Margaret Thatcher and her commissioner, Lord Cockfield, was rolled out as the '1992 Programme' and had the Anglo-Saxon model of liberal markets at its core. It was promoted as a natural extension to take these ideas forward into energy markets. The early debates were all about access by competitors to the electricity and gas networks. Competing supplies needed level playing fields and dominant incumbents could not be trusted to provide these. Hence the Commission pushed for *regulated* third-party access (rTPA) to the electricity and gas networks. For the national champions this was anathema. It meant a loss of control and undermined the vertically integrated model upon which they had relied for much of the twentieth century. They wanted *negotiated* third-party access (nTPA), and of course they would be doing the negotiations.

Other factors were to undermine the big players much later, but back in the 1990s they fought a defensive ditch-by-ditch game with considerable skill and against an under-resourced Commission, which had at best wavering support from national ministers. It took three rounds of directives to put the legal and regulatory framework in place, and although many on the Commission side counted these legal changes as ultimately the determining causes of liberalization, in fact the break-up of the dominant companies and the development of limited competition are best explained by the general excess supply and changes in IT and financial markets.

It has taken more than a quarter of a century to get most Member States to comply with the directives, and in particular the Third Energy Package, which was finally adopted in 2009.[4] The companies responded by engaging in the largest mergers and acquisitions (M&A) spree in their history, in effect creating pan-European champions. At the forefront were RWE and EDF, with supporting roles from Vattenfall and ENEL. The other big player, E.ON, was formed from 2000 onwards from a series of mergers, including the acquisition of the German gas networks (Ruhrgas). It moved to the top slot in size before being felled by the consequences of German domestic energy policies, which then led to its break-up less than two decades later.

The takeovers did create pan-European giants, and as each entered each other's market they created a pan-European oligopoly. The Commission's attempts to get a competitive market going were consequently hampered by the fact that there were now far fewer players on the field. The Commission convinced itself that if each entered each other's market, this increased competition in those markets, and since European networks were poorly linked, it did not matter that in the process they eliminated potential entrants in their own markets.[5]

The war of attrition led to the recognition that only by separating out the core networks in electricity and gas would competitive access be possible. Unbundling became the next battleground, and this turned on *legal* unbundling versus *ownership* unbundling. The companies won a Pyrrhic victory and were able to maintain their vertical structures, but this was overshadowed by the coming of infrastructure investors focused on regulated assets and asset bases, and with very different costs of capital. The result was that the companies divested themselves of their networks for financial reasons. Even France's giant EDF has had to consider this path. It was the first breach of the vertically integrated structures. More were to come.

The big prize for the Commission and the advocates of competitive markets was not, however, the instrumental one of breaking up the dominant companies, but rather finding a way of getting customers to be able to switch suppliers and thereby exercise the power of choice. Liberalization in theory was what the directives forced through, but there is a world of difference between the *right* to make a choice and the *ability* to do so in a straightforward and hassle-free way. This would take the coming of new technology, and eventually smart meters. But in the meantime the overwhelming feature of European electricity and gas markets is that most customers do not switch. Competition is restricted largely to a game between the incumbents, with a fringe of small entrants, and, as the British example has shown, this leaves a lot of market power with the incumbents.[6] This is the structure that now has to face up to our predictable surprises and, as we shall see in Part Three, it is not well placed to succeed. The incumbents may be making hay now, but their time in the sun is drawing to a close.

The completion of the internal energy market was a project conceived in the decade after the fall of the Berlin Wall, before climate change came onto the energy policy scene. There were, therefore, no other policy objectives to get in the way. It was in this sense a simple exercise: in the context of abundant supplies and post-1986 lower fossil fuel prices, the policy objective was cheaper and more competitive European energy, and the means was competition. Governments did not have to worry much about the energy mix – particularly its carbon characteristics – and security of supply was not on the agenda, not least because deindustrialization was reducing demand, and hence investment in new energy assets was not a priority.[7]

The climate change agenda

Much of this would change in the second decade of trying to complete the internal market after 2000. The climate change agenda was one that the Europeans were keen to champion. The disconnect between European citizens and the Commission had grown wider, and climate change offered two key political attractions: it could be a European rather than a national policy (even the British saw this); and it could appeal to the young and connect them to the European institutions. The emergence of 'green' voters necessitated moves by the dominant political parties to take account of climate change and to try to integrate these voters into mainstream political structures. Better still, the Germans could take the lead, and the British could sign up too.

The German environmental stance was conditioned by the nuclear issue, and in particular nuclear weapons. Germany had been on the front line of the Cold War since the Second World War, and as the superpowers (and Britain and France) lined up their nuclear weapons against the Russians, it could expect to be annihilated in any nuclear war. Worse, it was a divided country, physically straddling the East–West frontier.

The anti-nuclear position did not stop Germany developing civil nuclear power stations, or indeed even Helmut Schmidt asking the US to deploy cruise missiles on West German soil.[8] But much changed as the Soviet Union began to fracture. The Chernobyl nuclear accident in 1986 was a key moment when the German Green Party began to gain traction, and after the Soviet Union collapsed, the military imperative and the threat of invasion withered away. With the Soviet Union gone, and with reunification, Germans could afford to become anti-nuclear.

German Greens advocated a decentralized local renewables electricity system without nuclear power stations. In 2000 the Green Party made its first breakthrough, able to form a coalition with the Social Democrats (SPD). The price for the coalition was an announcement of a phase-out, with the early symbolic closure of one nuclear power plant.[9] It would take the 2011 Fukushima nuclear disaster to turn this into a rout of nuclear, with the immediate closure of eight nuclear power stations[10] and the complete closure of all planned by 2023. Thus was formalized the *Energiewende*, though its origins were earlier.

There had to be an alternative, and for the Greens that was a combination of wind, solar and energy efficiency. Germany was the driving force behind the European Climate Package, negotiated in 2008 and formally adopted in 2009. Europe signed up to Kyoto and adopted a set of targets, all with the magic number '20'. By 2020, emissions would be reduced by 20%, and there would be 20% of energy from renewables and 20% gains from energy efficiency. Nuclear was of course not designated as a 'renewable': Greens were at least as much anti-nuclear as they were in favour of reducing emissions. Indeed, in Germany, nuclear would give way to more coal and, remarkably, by 2015 German emissions were actually rising (as shown in Chapter 2). In practice the *Energiewende* turned out to involve a shift from nuclear to coal (and for other reasons from gas to coal).

The 2020-20-20 package had a neat political ring to it, but it was economically illiterate. The chances that the numbers would all add up to the magic number 20 were zero.[11] But in one sense this did not matter. For as we saw in Chapter 1, most European politicians were convinced that the oil and gas prices would keep going up in the next decade, and that therefore what they had on their hands was a clever, globally competitive strategy which would give Europe relatively cheap energy when set against what they assumed would be much higher fossil fuel prices confronting their competitors (the US and China). They were blissfully unaware of the shale revolution, and convinced they knew the future.

The German *Energiewende* was a national policy decision. There was nothing European about it (just as there would be nothing European about Merkel's other dramatic and decisive moment – opening German borders to the refugees from Syria and the Middle East in 2015). No one else was consulted. Fukushima happened in the run-up to a crucial regional election in Baden-Württemberg, and this clearly made for a panicked decision about the future of nuclear as the Japanese accident played out all over the media.[12] Germany's comparatively dark skies would be the energy source for first-generation solar panels on a massive scale, together with wind, costing the German consumers around €20 billion in subsidies per annum.[13]

With the *Energiewende* in place, it was in Germany's interests to foist the same costs on the rest of its European competitors, especially once the falls in fossil fuel prices began to expose the fallacy of assuming that the oil and gas prices would come to the rescue. Germany's very dirty coal secret (discussed in Chapter 2) is also something that will have to be tackled, and it has only wind farms and solar panels with which to respond. Europe's energy future, as far as it is driven by Germany, would be based on these renewables. To the extent that others take a different view, policies will be less European and more national.

Security of supply, Eastern Europe and the Turkish connections

Other EU member states are in a different place, both politically and geographically. The biggest difference is with the Eastern European former Soviet Union countries which joined the EU in 2004, notably Poland, Hungary, the Czech Republic and Slovakia. These countries' political memories are conditioned by the long years of Soviet, and hence Russian, repression. For them, the Second World War effectively ended in 1989, not 1945. Energy security was part of the Soviet control mechanism, and the countries were tied into high-level grids and Russian dependence. The ability to literally turn out the lights is a threat which means much more to them than it does to the big Western countries; subsequently, security of supply and moving away from Russia and, in particular, Russian gas dependency, is a priority. For Poland, this means domestic coal, and then LNG gas and nuclear. For the Baltics, with almost complete Russian dependency initially, particularly in Estonia, it means connections to the West. For Hungary and Slovakia it is a game of influence and control between elites and companies, and a recognition of the close proximity of Ukraine, and of Russia's designs on it.

Two relationships have shaped the more recent debate about European dependency on Russia in the context of rising gas prices. The first of these is with Ukraine, which serves as the main pipeline artery delivering Russian gas to Europe (as shown in Figure 6.2). Ukraine is a big energy hub, with not just the pipelines but the storage too. It is also a big gas and coal user, notably in its eastern Donbas region.

What few Europeans appreciate about the tangled relationship between Russia and Ukraine is that the latter used to literally be Russia. As noted in Chapter 6, the Rus people started out there, with Kiev as their capital. Yet, as few Russians appreciate, western Ukraine has been part of Europe, and the Austro-Hungarian Empire in particular. Northern and north-west Ukraine was part of Poland at one stage. Ukraine has always been under siege, and its identity is defined by these competing claims.

Throughout most of Putin's rule, getting Ukraine back into the Russian sphere of influence has been a core strategic objective. The interruptions in gas supplies in 2006 and 2009 need to be seen in this context.[14] They were not merely the result of narrow disputes about payments, and they were never discrete events. They are part of a long game, to be seen in the context of the annexation of Crimea and the civil war in the eastern provinces. The tactics, such as the interruptions, are opportunistic, but the end is not.

The Europeans tried to keep the energy issues with Russia in a separate security box, at least until the Crimean annexation, which led to more general sanctions and a souring of relationships with Russia. Putin calculated that the threat to Europe's gas supplies would encourage the Europeans to support alternative pipelines, and in particular his grand strategy of Nord Stream and South Stream to bypass Ukraine. As noted in Chapter 6, and in line with Putin's expectations, the companies fell over themselves to support both, with the politicians struggling to keep their distance. Nord Stream 1 was built, but eventually the Commission had the courage to see off South Stream.[15]

Success eluded the Europeans when it came to securing alternative supplies of gas from the Caspian Sea and via the Turkish route from the Middle East. The Nabucco project would have provided a way of bypassing Russia. The US had already succeeded with the Baku–Tbilisi–Ceyhan oil pipeline, but only once the Americans had declared it to be of strategic importance and hence provided Azerbaijan with political protection in defying Russia. An EU commitment did not have the same clout, and in any event the Europeans did not have special trade and other opportunities to offer in exchange or – ultimately unlike the US – the military forces to back it up. The Nabucco pipeline would have gone through Turkey, and here the Europeans continue to have a tricky relationship which goes back to the Ottomans and earlier.

The relationship with Turkey has been a fraught one. Turkey was offered the prospect of EU membership, but never with much sincerity. Several key European countries would find it very difficult if and when it actually came to a decision to accept Turkey. So instead a game has been played out, with membership negotiations conveniently running into the long grass over Cyprus. It has only been the refugee crisis of 2015 and 2016 that has unblocked some of this, with Germany in particular trying to induce the Turks into stopping the flow into Greece in exchange for concessions on visas and the accession processes.[16] But EU membership looks even less likely.

This was not, however, the end for the Southern Corridor bypassing Russia. The Trans-Adriatic Pipeline (TAP) is going ahead, connecting the Trans-Anatolian Pipeline (TANAP) to the Shah Deniz II Caspian gas field. This goes close to Turkmenistan, Iran and Northern Iraq, opening up the possible eventual European access to their gas fields too. Ultimately it may turn out to be a tussle for supplies with China's Silk Road pipelines. The fact that the alternative – Turk Stream – has faltered with the problems between Turkey and Russia over Syria helps the TAP/TANAP case. These difficulties return us to the long history of the Russia–Byzantine–Ottoman conflicts of the previous thousand years.

Other options for diversifying supplies included Libya, but these too ran into difficulties with the Arab Spring and subsequent chaos as Gaddafi was toppled from power and then killed. For the Southern European countries, and especially Italy, this cut off another diversification option for gas.

There was one alternative gas option in the short term – LNG. This was developed before the prospects of shale gas were fully understood, and in particular the impact of the US ceasing to be a gas importer. Europe sprouted lots and lots of LNG terminals (as shown in Figure 8.3) – investments that

Figure 8.2 Main Southern Corridor pipelines

Source: Aurora Energy Research

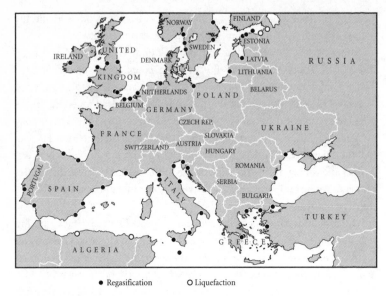

● Regasification   ○ Liquefaction

Figure 8.3 Key European LNG terminals

Source: Aurora Energy Research

put the pressure on Russia and limited its abilities to interrupt supplies, albeit at high cost. This was the one investment that would turn out, for reasons not envisaged at the outset, to be a good long-term bet. Europe has ended up with enough LNG capacity to meet nearly half its demand for gas.

The final option to escape Russia's gas bear hug is to get out of gas altogether. In the narrative of European energy policy, the threat from Russia morphed into a concern to get out of gas more generally, and this in turn fed into support for three different interests. The French saw this as an argument for new nuclear and the rolling-out of the European pressurized reactor (EPR) across Europe, starting with Finland, then France and on to Britain. The German Greens saw this as an argument for renewables. Finally, the Poles (and initially the Germans and British) saw it as a reason for hanging on to coal, and indeed in the mid-2000s Britain planned more new coal (including Kingsnorth power station), whilst Germany was actually building almost 13 GW of new large coal power stations.[17] It was 'anything but gas'.

The central power of Germany played a pivotal role in adjudicating between the three objectives of energy policy that emerged from the internal market, the climate change challenges and the security concerns. For as long as Poland and the new Eastern European members were focused on anchoring their countries in the Western political and military structures (in practice the EU and NATO), many were willing to toe the German line. They even signed up to the Kyoto targets and the Renewables Directive. So far, so German.

But as time has passed, the memories of the Soviet occupations have faded. It is over a quarter of a century since the Berlin Wall fell, and only those over the age of forty have any real memory of its nastier features. But whilst time faded the memory, other more threatening challenges emerged, notably the economic crisis that began in 2007–08. Before then, the economic boom after the fall of the Soviet Union and the urgency of tying into Western institutions had meant compliance with the internal market, and even the 2020 climate change package. But then came the euro crisis, mass unemployment and an end to their dream of immediate European prosperity. By 2015, the impact on Eastern European politics was being felt, with new, much more nationalistic (and typically right-wing) populists taking charge. Add in the migration crisis and the possibility of economic collapse in Greece, and suddenly energy was a matter of national security, and in Poland jobs for coal miners, with climate receding into the background.

The Paris summit in December 2015 saw Europe united in principle but divided in practice. Europe could adopt *European* targets, but the detail of their disaggregation into *national* ones remained to be determined. Poland again led the resistance. The idea of European climate leadership had been easy in the 1990s, as the rust-bucket industries of the East collapsed, and as Europe gradually deindustrialized, substituting carbon production, now in China, for carbon consumption of Chinese energy-intensive goods. But it turned out to be more illusion than hard constraint, except in Germany (and to an extent in Britain) where it had some political edge. If more carbon constraints imply higher costs and less security, future policies will be resisted much more emphatically, particularly in Eastern Europe.

Taking stock: Europe faces the new energy world

Europe faces its energy future in poor shape. It has few natural resources. North Sea production is running down in the British sector. Fracking has been kept largely off the political agenda by the Greens and NGOs. The nuclear renaissance has gone backwards. The Finnish reactor is an economic disaster for its French developers, bringing down Areva in the process. Flamanville in France is late, over-budget and has been subject to lots of technical issues. Hinkley in Britain has become more controversial as the project has developed. France in the shape of EDF faces enormous financial challenges: EDF has been borrowing to pay its dividend in recent years – in effect a form of finance for the state's budget – and as a result has large debts. And then EDF needs to pay for the upgrading of its existing nuclear fleet, provide for decommissioning and bail out Areva, before paying its share of the Hinkley project.[18] On top of this are the further costs in addition to its existing fund for decommissioning. Other nuclear options are also on the table, including a Russian-made nuclear reactor in Finland. But these all come with considerable problems, and hence limit the scope for a future nuclear renaissance across Europe.

EDF, like its counterparts in Germany, RWE and E.ON, faces falling wholesale prices. The financial consequences have already been dreadful, threatening their future viability. Some could conceivably go bust in due course, though they are more likely to suffer a slow death. If national champions are relied upon to come to Europe's energy rescue, there will be lots of disappointments. As we shall see in Chapter 10, things can probably only get worse for these conventional vertically integrated big electric utilities.

This is all about the internal problems, but they have a further alarming external perspective. Europe's bet on rising oil and gas prices has turned out to be wrong. Instead of marching towards 2020 with lots of increasingly competitive renewables, and out-competing the US (and China) as a result, the end of the commodity super-cycle and the coming of US shale gas mean that Europe no longer has much energy-intensive industry left, and it is not reshoring manufacturing from China. Instead it is the US that is buoyant and investing in more energy-intensive businesses.

On climate change, the Europeans effectively ceded the leadership at the global level to the US and China at Paris. The Europeans' plan had been to build on Kyoto to provide a further top-down legally binding set of targets. That was their pitch at Durban. At Paris, the Kyoto 2 model was replaced with an altogether different structure, based on voluntary pledges and derived from the relatively weak commitments in the US–China climate agreement ahead of Paris (which Trump subsequently reneged on), as we saw in Chapter 2. Going forward, globally, few look to Europe as an example to emulate. It has not demonstrated a successful low-cost route to decarbonization. The European exit from nuclear and its embracing of coal, combined with its loss of global competitiveness, are not viewed elsewhere with quite the enthusiasm of some European politicians. It is late in the day to be considering a reverse gear, and probably not politically feasible for many European countries.

On security, the Europeans have not managed to break free of the Russian bear hug (even if they have been using less gas), and they are not much nearer than they were a decade ago to gaining significant independent access to the Caspian Sea and via Turkey to Middle Eastern supplies. The renewables have rendered the power systems less stable, with the intermittency from wind and solar creating the need for back-up supplies, and much more generation capacity for any particular load. This in turn has rendered the conventional power stations interruptible when more wind and solar is available, exceeding the total demand. Making gas-fired power stations intermittent undermines their economics, and they therefore now also need to contract for gas supplies on an interruptible basis. Across Europe this large-scale generation source has been adversely affected even as gas prices fall.

Taken together, Europe is failing on all three main objectives. Its energy is relatively expensive, especially in comparison with the US; it lacks security; and it is no longer leading on climate change. Yet much of this is about to change. It is about to get a bit luckier, even if not as lucky as the US.

All three predictable surprises coming to the rescue

When oil and gas prices were high, mainland Europe transferred wealth from its consumers and businesses to the producers – to OPEC for its oil, and to Russia for its gas. (The exceptions are Britain and Norway.) The difference between the production costs of oil and the price, and similarly for gas, went as economic rents to the producer countries. The reversal of this massive wealth transfer matters. The impact of the price falls is, for example, bigger than the stimuli from fiscal and monetary policies after the Great Recession, including all the QE. It comes through a series of transmission mechanisms. Consumers have more money to spend on other things, companies have lower production costs, and the trade balances improve too. Consuming countries gain, producers lose. For Europe this is unambiguously a good thing for all but Britain and Norway (as oil and gas producers), and even Britain may be a net gainer as its oil and gas output declines. Economic growth returned in 2015 and 2016 partly as a result.

Whether or not the prices go back up again in the short term, Europe faces a relatively benign medium-to-longer-term energy outlook, and this should increase demand and hence economic growth. This gain will come not only from the oil price, but also indirectly in terms of security from the weakening impact of the end of the commodity super-cycle on Russia. With Russia having undermined its reputation for delivering security of supply for gas by its actions in interrupting Ukrainian supplies, and then its reputation as a reliable partner in geopolitical terms by its annexation of Crimea and the instigation of civil war in Ukraine, Europe has reacted by promoting renewables and even nuclear, and by building lots of LNG as described above. Its security has thereby been considerably enhanced, and Russia's ability to threaten Europe's supplies reduced.

Now that US shale gas is coming onto a market already saturated by the surpluses coming out of Qatar and Australia, which were destined for the US and Chinese markets, respectively, and now that the demand from the US has fallen away, Europe can take up the slack. LNG represents a threat to the Russian pipeline gas supplies. The only way that Russia can deal with this is to cut its prices and try to force the LNG out, analogous to the Saudi attempt to push out US shale oil. To signal its intent, Gazprom is promoting a second Nord Stream pipeline, despite the fact that Nord Stream 1 is not yet fully utilized. The obvious plan with Nord Stream 2 is to threaten to flood the European market, and to kill off Ukraine's revenues in the process. Regardless of whether it succeeds, the strategy will work only if it can be sustained and if the demand for gas holds up. On the former, Russia would have to live without foregone revenues for quite a while; on the latter, the future is in the balance. In this game of strategy, it is not obvious that Gazprom will win.

The real question is whether Europe will be using these cheaper fossil fuels at all. Lower fossil fuel prices mean that the relative price of renewables and nuclear has gone up. Whilst those already on the system, with near-zero marginal costs, can still economically operate faced with lower rival fossil fuel prices, this is not necessarily true of new investments. It also applies to the take-up of electric cars. Europe plans to significantly phase out fossil fuels by the middle of the century, so its gains from the lower prices will be limited if the objective of decarbonization is met.

With lower fossil fuel prices putting more competitive pressure on renewables, and with the mistakes made in committing so many resources to first-generation renewables, a binding carbon constraint (our second predictable surprise) might at first glance look like bad news for the Europeans. Higher temperatures have considerable costs at the global level, but the impacts are not evenly distributed. For the Europeans, warming by up to 2°C will probably do little harm to the main economies.[19] Agriculture is no longer an important industry in terms of its share of GDP, and for many areas warmer winters, earlier springs and longer growing seasons are good news, especially if accompanied by lots of rainfall. For the rest of the economy, warmer winters mean cheaper heating bills and lower cold weather emissions which can be particularly damaging.

These relatively small impacts depend on the climate staying at or below the 2°C warming. After that, all bets are off. In any event, Europe is responsible for quite a lot of the carbon stock in the atmosphere and can and should have a responsibility to deal with the consequences.

Hence, whatever its short-term costs and benefits, the carbon constraint will bite. What matters then is how Europe decarbonizes. With so much intermittent wind and solar already on European systems, new technological solutions are required. One mistake throws up another challenge. For renewables the requirement is for storage, and hence to harvest what is so often wasted. Because wind is intermittent it not only causes intermittency in other forms of energy generation, it also necessitates the need for the demand side to be intermittent too. That is what demand-side response mechanisms are for. Then there is the new market for smaller peaking plant and the decentralization of electric systems.

The coming of the new technologies and the transformation across the full energy supply chains may have arrived just in time for Europe, and will first mitigate some of the negative impacts of the dash for first-generation renewables, and then offer considerable upside. Next-generation solar has quite a lot of infrastructure to build on. Smart grids and smart meters make the grid and distribution networks better able to absorb the variance from intermittency and operate more decentralized systems. Electric cars can be charged in a way that takes account of the times of solar and wind surpluses. New batteries and other storage options help to balance the intermittency. All of these drive down the costs of decarbonization, stabilize the energy networks and, in the process, further reduce the dependency on fossil fuels, even as these get cheaper.

As discussed in Chapter 3, the impact of new technologies goes well beyond the changes in the electricity and energy supply chains. The digitalization of the economy as a whole, from robots and 3D printing to AI, not only increases the importance of electricity, but also plays to Europe's strengths. As with the US, these technologies rely on deep intellectual capital, high levels of education, and the scope for innovation and entrepreneurship. The US dominates Europe on entrepreneurship, but both economies are supported by liberal open democratic institutions and freedom of thought, expression and ideas. Although some of the new information technologies may infringe on individual freedom and privacy, these core strengths remain. They contrast sharply with China (and Russia too).

A further benefit for Europe from the new technologies is locational. As argued in Chapter 3, the need for outsourcing and the role of labour costs decline in the new digital world. Thus, products tend to be produced closer to customers. The ideas can come from anywhere, but translating them into products and services is typically more local.

The key point about this digital world is that the cost of labour becomes less and less important. Europe's Achilles heel in manufacturing is therefore much less important in the new technology world. There is a shift from labour to capital. For Europe the problem that arises is what to do about the resulting income inequality and the diminished prospects of the less educated and skilled, and those without capital. The European welfare state model probably cannot withstand these pressures on its own, and hence more radical ideas, like the provision of universal basic income, may then come into play.

Finally, a strong distinction between the US and Europe is the extent to which politics can produce big swings away from economic efficiency and inhibit markets. In the US, the shale revolution worked because it was allowed to, in response to the price signals. The technologies came together, and the industry could develop and respond fast, both up and now down. The switch from coal to gas took place quickly because market signals worked. Indeed, what is interesting about US energy policy is the way in which it tends to follow, rather than lead, market developments. The current examples of attempts to regulate coal out of the market follow the falls in gas prices. It would probably happen anyway, and the US regulatory approach may, post-Trump, follow the trend. It is of course not always so – for example, with the CAFE regulation of standards for vehicles. But even here the actual rules followed the market pretty closely.

In Europe, political ambition plays into a much greater belief in the power of the state and the ability of politicians to make decisions superior to those of markets, and indeed to be able to predict the future. The *Energiewende* and the EU 2020 climate package are the two key examples. The political culture in Europe is very different. When it comes to energy, this European bias towards intervention has not worked well in the past. This is most notable in climate policy and, as noted in Chapter 4, the US has done just as well without all the policy costs.

Some state intervention is necessary in the R&D space, but again the Europeans and the US differ. Whilst the US does a lot of state-driven R&D, the big breakthroughs from this side come from defence. This is where the state engages in developing ideas like the Internet and communications technologies, new materials, nuclear expertise and now robotics and AI. New forms of digitalized warfare are a big stimulus to R&D in the US. The space programme resulted in lots of R&D and in both space and defence there was a plethora of spin-offs. In Europe, outside Britain and France, defence is not a major player, and instead R&D support comes mainly through the universities.

Here again the US has an edge. It has many great universities, and Europe has few outside Britain in the global top 10 or even 100. But the US has one further R&D advantage: it has the business culture to innovate and create new businesses with core R&D. It is hard to imagine that the great Internet and communications companies in the US could have been European; those European companies that tried, like Nokia and Siemens, failed to become global players in these areas.

As the technological predictable surprises unfold, Europe may not have the lead its intellectual and open societies should give it. European leaders like to pick winners rather than ideas, and they fail to concentrate instead on supporting R&D and innovation. Europe may be lucky in that the weight of its dependence on fossil fuels has been lightened by the new abundance and the lower prices, but it remains to be seen whether it exploits the opportunities that the new technologies offer.

PART THREE

Creative Destruction and the Changing Corporate Landscape

Falling oil prices in the medium-to-longer term, and the gradual exit from fossil fuels, will eventually be an existential threat to the great twentieth-century oil and gas companies. Their products will no longer have mass markets. Not only will demand fall, but in the process they will be the target of the green NGOs and climate change campaigners.

Indeed, it has already happened. New targeted campaigns have started to claim that their assets will be stranded and to challenge their social licence to operate, especially when they are developing new reserves deemed by their protagonists to be pushing beyond the total carbon budget consistent with the 2°C limit.

As businesses, what should they do? Should they go 'beyond petroleum', as BP tried and failed to do? Or should they just tough it out, with the expectation that the 2°C will be breached anyway, and the knowledge that other companies, especially state-owned ones, will carry on regardless? Should they harvest and exit, paying out rising dividends over time, and stop investing in new reserves?

Whilst the plight of the oil and gas companies is at least a well-defined problem, that of the electric utilities is rather different. Since electricity is the likely big gainer this century, it might seem a good place to be. But the electric utilities are structured around a very particular – and very twentieth-century – model. In a world of no storage, the need for the instant matching of supply and demand has necessitated the sorts of vertically inte-grated and big electricity companies that emerged. The fact that the demand side was passive, and small-scale flexible generation has been expensive, simply reinforces this model. It is a world of monopoly and vertical integration where size matters.

All this is changing and, in some countries, changing fast. The new technologies are overwhelmingly zero marginal cost, creating a world of capacity rather than energy-only markets. Flexible generation, smart grids and smart meters, electric cars, batteries and storage are changing the game. The electric utilities are well designed structurally to address yesterday's markets. They look less adapted to this technological future.

What should the electric utilities do? Should they try to become renewables and small-scale flexible generation operators? Should they abandon the large baseload power station model of big nuclear, big coal and big gas? Or should they harvest and very gradually lose market share? Should they get into transport?

The chances are that both big oil and gas and big vertically integrated utilities will die, their economic relevance slowly fading away. But none of this is inevitable. There are choices, and some choices are much better than others. The history of corporate strategy and behaviour signals a pessimistic answer. Few companies faced with radical technical change have made a big transition. History is littered with companies that fail to change and gradually die, left as ghosts of their former selves. But some do survive and prosper.

In Part Three we look at the plight of the big players, before turning to the new market designs that are emerging in the digitalized, zero marginal cost world – and who is likely to prosper in this very different emerging context.

CHAPTER 9

The gradual end of Big Oil

Think of oil and what comes to mind are names like Exxon, Chevron, BP and Shell – the Seven Sisters and their successor companies we met in earlier chapters. These companies stood astride the twentieth century and their oil powered the greatest economic transformation in the history of the world. They controlled most of the world's supplies up until the end of the 1970s, when the NOCs took back both ownership and control of their resources. Into the 2010s they remain among the largest companies in the world (though dwarfed by the NOCs like Saudi Aramco),[1] comprising a big chunk of the US and UK stock markets. Their dividends pay the pensions of millions, and they account for a great deal of employment and tax. They dominate because they are big; and they are big because their vertically integrated model has been a winning one throughout all the twists and turns of the last century.

Prior success does not guarantee future prosperity. Few companies have ever escaped the decline of their prime activity and morphed into something else. Coal companies did not generally become oil or electricity companies. Rail companies did not dominate the motor industry, and few carmakers have built large aviation businesses.[2] Oil companies can (and have) become big gas players, but that is because oil and gas tend to go together in the same sorts of wells and using the same sorts of technology. It is hard to get oil without getting gas too. Oil and gas companies have not become electricity players, and their forays into gas power stations and renewables have been at best unimpressive. Their attempts to go 'beyond petroleum' have largely failed, though this has not yet stopped them trying.

The price falls at the end of 2014 and into 2015 have been a setback to the business models of these companies, but this may be just the first blow in a longer and mostly gradual decline. The impact on their share prices and their valuations has been brutal, as Figure 9.1 shows.

They have reacted as they did in the mid-1980s by cutting back on capital expenditure, reducing operating costs and selling off marginal assets. This time they have had to start borrowing to pay their dividends too. Though they survived the earlier price falls in the 1980s and 1990s, this time may not be so easy, and their years of growth may be finally coming to an end. They may not be on the roll call of the success stories of this century as they were of the last.

To understand why their strategic options are so narrow and why the odds are against them, it is necessary to first consider their strengths and

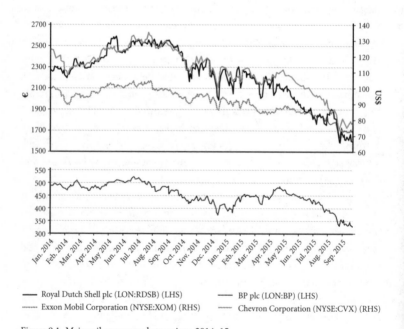

—— Royal Dutch Shell plc (LON:RDSB) (LHS)   ------- BP plc (LON:BP) (LHS)
········ Exxon Mobil Corporation (NYSE:XOM) (RHS)   —— Chevron Corporation (NYSE:CVX) (RHS)

Figure 9.1 Major oil company share prices, 2014–15

Source: Google Finance, 2015, https://www.google.co.uk/finance

what made them so dominant in the twentieth century. This creates a platform on which to consider how they might be undermined by the three predictable surprises discussed in Part One: by falling long-run prices, by decarbonization, and by new competing technologies.

The twentieth-century model – and why it worked

Why did – and do – big oil and gas companies dominate? There are two main explanations: first, the cost structures and nature of the risks have meant that ever-greater scale has been economically efficient; and second, market power enabled these companies to keep out rivals and exploit their customers. Big projects need deep pockets and skills, and smaller players and entrants do not have the resources to resist when the squeezes are put on them. It turns out that both characterize the oil and gas industries, and that they are related.

Oil has several characteristics that, whilst not unique, together define a particular industrial structure. The full vertical chain starts with E&P – looking for new deposits. This is a risky activity: it is a (often literally) sunk cost since if it fails the costs cannot be recovered, and there are bound to be failures, lots of them. Prospecting geologists do not themselves earn revenues. Nor do the hired hands among lobbyists and diplomats, trying to get access to reserves in countries that might not share the aims and ambitions of Western consumers. It took deep pockets and political clout as well as bribery, corruption, mavericks and inside information to work out where to drill and who to pay in a pre-IT world. The early contractual arrangements with the Middle Eastern countries (and Venezuela and Mexico in Latin America) proved to be at best non-transparent, and often to involve not just outright corruption but even military force. The overthrow of Mosaddegh in Iran in the 1950s, and the consequent reversal of nationalization, is just one among many examples we have already noted. These elements of corruption, bribery and occasional force were the necessary ingredients for the companies to do business.

Once geologists have found likely prospects, test drilling can start in order to discover if there really is oil and gas in sufficient abundance to merit the required extraction infrastructure investment. Many wells turn out to be dry or simply not worthwhile. Shell's recent foray into the Arctic cost over $2 billion, a particularly expensive write-off.[3] Pre-IT, this was sometimes as much a matter of art and luck as science. Knowledge comes from geologists but also experience, and this has often been built up within companies. Sometimes it makes sense to allow small entrepreneurial companies to take the risks, and then to buy them out, much as in pharmaceuticals. This has worked well in US shale oil and gas. At other times, in-house teams built up in depth are the answer.

Finding the oil is only the start. The next step is to invest in the wellhead infrastructure. This is a highly capital-intensive activity, and it takes time to get from the initial drilling to full flow. This may go on for many years before the field is depleted, and then only after more spending on trying to keep the pressure up in the wells. As noted earlier, oil fields are typically abandoned before half the oil is abstracted.

The traditional oil and gas development therefore involves a large fixed and sunk cost. It is a risky business, made more so because future prices are hard to hedge: futures markets are even now relatively short-term, and rapidly become thin. The companies therefore have to take the risk themselves, and by having lots of different projects – a portfolio – they can do a bit of internal risk-spreading. Only the biggest can contemplate the Arctic, the northern Caspian, the Alberta tar sands or offshore Brazil, and even then they often need to share out the investments with the host NOCs.

Yet this is only part of the capital costs of the supply chain. Refineries are also large fixed- and sunk-cost investments, as are pipelines and retail distribution networks. The sheer scale of the continuing capital requirements puts the industry in a class of its own – probably rivalled only by shipbuilders and manufacturers of heavy military weapons and their supporting infrastructures. They require access to capital and financial markets – access to the capabilities of absorbing, managing and spreading the investors' risks.

Large oil companies with a portfolio of E&P, full production fields and declining fields are able to deliver more stable financial flows, and can use some of their retained earnings to finance the development of future production. It is an industry that operates on a pay-as-you-go basis: it uses its current earnings and its current balance sheet to finance current investments. Each generation pays for the next and, as a result, it has typically had low borrowings (though this may now be changing to prop up dividends in the face of the falling oil and gas prices). Start-ups and project-only companies are not like this: they are pay-when-delivered, and they need investors to concentrate their exposures on their specific projects. The cost-of-capital differences between these two models are enormously favourable to large integrated oil companies. That is one reason why they exist and have prospered as a preferred model.

The central role of market power

A key problem with the long time horizons for the recovery of the fixed and sunk costs is that the oil price fluctuates. Investors not only need to know that the risks are diversified across a big portfolio (so that they were not betting on a 'one-horse race') but also that, in taking long-term bets, the companies can weather price fluctuations. In the post-financial liberalization era, and with the growth of instruments to hedge these sorts of risks, capital markets can do some of this job. But even now there really is not much protection for the price risks, which might stretch over two or more decades, and in particular from the sorts of price falls witnessed from late 2014.

There were essentially two ways this price risk could be handled in the twentieth century. The companies could 'manage the markets' through collusion and cartels, or governments could take the risks through stakes and even outright ownership. Of these, the companies went for the first option, fixing prices downstream. The ways they did this are legendary, starting with buccaneering capitalists and 'hate figures', like J. D. Rockefeller and J. P. Morgan, who used the most aggressive means available to them, revealing just how ruthless monopoly capitalism could be.[4]

The monopoly solution required control over each and every part of the vertical supply chain, with the refineries in the middle and the railroads and pipelines as the key choke points to eliminate rivals. Although history highlights the successes of the trustbusters associated with Theodore Roosevelt's Progressive Party, and in particular the break-up of Rockefeller's empire and the emergence of the Seven Sisters in the early part of the twentieth century, the companies generally won the battle to control prices, and managed the contracts and the trading to their mutual advantage. The successors of Rockefeller's Standard Oil went on to replicate much of his model. The monopoly may have been killed, but an effective cartel lived on.

Downstream market power was the means by which the huge fixed and sunk costs could be passed through to customers. What remained was to fix the upstream – the contracts with the producer countries. The Seven

Sisters did this through collusion. They standardized contracts around 'sharing arrangements' with producer countries, and then offset the tax back home. It worked extremely well for the companies, and the long period of low prices from the late 1940s through to the 1970s was a golden era of ever-greater supplies and very stable prices. They continued to make good profits in a world of very gradually falling prices.

Imposing upstream market power eventually begets both producer and competitor responses. As long as the Seven Sisters had a stranglehold on access to downstream markets, producers had little option but to fall into line. Who else could they sell to? There were no other significant markets: China remained isolated, and Japan was only beginning to grow – and, after the Second World War, was in any event solidly in the Western camp. Selling oil was all about access to the US, European and gradually the Japanese markets.

The producers matched the downstream market power by trying to create their own, and after more than two decades of trying, OPEC finally struck success in the 1970s. It could do this because there was little spare capacity, because the threat of diplomatic and military retaliation was weakened by the overarching Cold War, and because the Yom Kippur War created solidarity. All for one and one for all worked – for a while. Monopoly of upstream resources came to balance monopsony of downstream market power.

A second challenge to the Seven Sisters came from smaller rivals. If the producers could sell to their competitors, then perhaps they could gain greater control. Many countries promoted their own national champions to rival the dominant US and UK companies. Italy's ENI and France's Total were among this second tier, alongside individual entrepreneurs and smaller players. Yet these competitors could gain traction only if they had a degree of protection at home – protection for themselves against rivals, and government support in granting them privileged positions in the home markets, as they did in Italy and France.

This brings us to the role of politics and government. For as well as the large companies having intimate relationships with finance, they almost always had governmental support. BP and Shell could rely on the British government to assist them. The US government stood behind the successors to Standard Oil. Embassies and their diplomats saw their role as supporting their national flag-carriers, and military chiefs saw access to oil as a vital national interest. It was a symbiotic relationship: many oil companies fed information back to their home governments, playing the role of shadow and less formal diplomatic outposts.

An example is the British role in Iran, which stemmed from the need to gain oil for its navy. The hatred directed at the British in Iran after its revolution in 1979 can be traced back through the century. Britain and France created the country borders in the Middle East, and indeed the states of Iraq and Syria, now the subject of war and terror nearly a century later. This was not an open global marketplace, but rather a set of spheres of influence. Big Oil, Big Finance and Big Governments were all bound together. For a century it worked. It might pose as entrepreneurial capitalism, but monopolistic nationalistic capitalism is what it was.

The search for new reserves

This cosy tripartite relationship of companies, finance and governments began to fall apart after the OPEC shocks of the 1970s. The first big problem was that the companies were squeezed out of the core producer countries. They hadn't been allowed much of a role in Russia since the Bolshevik Revolution. But now they were forced out of their prime positions in the key OPEC countries. Their assets were gradually nationalized. Within a mere decade or two, around 90% of the world's oil reserves would be in government, rather than independent company, hands.

As a result the companies found themselves having to deal with the new and increasingly large NOCs that had taken control of what they had once taken for granted. The NOCs' power and influence grew slowly. The immediate effect was to change the terms of trade, but gradually they moved into more and more of the supply chain. As they did so, the companies were forced to strike deals to gain access, often more as operators than owners, and to look elsewhere for reserves. Oil services companies, without owning oil or even producing it, grew in this new environment, and now companies such as Halliburton, Baker Hughes and Schlumberger compete for this business. As these companies have developed their core competences, the oil companies contract out to them too. The complexity of these relationships was seen in the 2010 Deepwater Horizon disaster – Transocean operated BP's well, and Halliburton cased the wellhead supported by a host of companies.

The nationalizations meant that the bedrock of the value of the companies to investors – their real physical assets represented by the companies' booked reserves – was being cut away. They had to find more oil to book as proven reserves in order to provide the valuation against which finance would be forthcoming.

OPEC also directly challenged the national security of the consumer countries, threatening and sometimes actually interrupting supplies. This gave a new twist to the governmental support for finding alternative reserves to book, and to secure ownership and control.

The imperative for the companies if they were to stay in business was to find more oil and gas that could be proven and added to their asset base. The great diversification began, starting with Alaska, moving to the North Sea, and continuing right up to date with the new frontiers in deep offshore waters. Pre- and even post-shale, these included the Gulf of Mexico, Brazil's eastern coast and the Arctic. Only large companies have the scale, technological capabilities and access to finance to take on these sorts of challenges. In a world that had been shaken by the OPEC shocks, the oil companies not only opened a new chapter, but from the investors' perspective a bright and profitable one. In the process, they chose to get even bigger.

The ratio of reserves to production has been the obsession, to be reported at every company presentation and AGM. A glimpse at how important these ratios became was revealed in the great Shell reporting scandal in the 1990s, when the company overstated its reserves by more than 20%. The temptation for any chief executive and finance director has always been to tell investors a good story. The higher the cover of existing production and new possibilities, the more secure investors would be.

The companies were driven by this supply-side imperative: keep finding more oil to replace what was being used up. Although other mining and natural resource businesses shared this characteristic, none made it the driving force of their business models on quite the same scale. It explains why these companies required the sorts of management with a longer-term perspective, and in particular why they tended to be led by powerful individuals with long periods at the top. The oil industry is littered with 'big men' – from Rockefeller and Nobel, through to Marcus Samuel (who founded Shell) and up to Rex Tillerson and John Browne, who shaped the modern Exxon and BP respectively. These are the sorts of industrialists who have biographies written about them, who tended to spend their lives working inside their particular companies, and who looked decades ahead. The current industry leaders carry on the search for new reserves, but face a world where it is far from clear that the demand their predecessors took for granted will actually materialize.

Most of these executives saw (and continue to see) themselves as very much at the cutting edge of capitalism, though the reality has always been rather different. They needed political skills to negotiate what was and is a very political industry. They needed to persuade their home country politicians to look after their interests abroad, and inevitably most were and remain de facto national champions. The big IOCs are still American, British, French, Italian and so on, and though quintessentially multinational in scope and activities, they remain very much national in character. The relationship worked both ways: politicians needed to make sure they had secure supplies to back up their military security, to keep an increasingly oil-dependent industry going, and to make sure voters had cheap and abundant fuel for their cars. The companies needed their home governments to lobby on their behalf, and the governments needed to have secure supplies.

The final part of this symbiosis that has evolved is all about the money. Whilst producer countries are heavily dependent on the royalties and licensing fees, as we saw in Part Two, consumer countries are addicted to oil taxation. In many Western countries more than half the cost of fuel to motorists is tax. Fuel is a classic good in inelastic supply: high tax does not choke off demand.

An example: BP

BP is a good example of the way in which these companies have developed. It was founded as the Anglo-Persian Oil Company (renamed the Anglo-Iranian Oil Company in 1935 and later BP in 1954), driven by the military imperatives of the British government (and Winston Churchill in particular) to secure a safe and reliable oil supply for the Royal Navy as it switched from coal to oil. Its early development depended on the British government's ability to persuade Iran, both militarily and diplomatically, to allow the production to carry on, and not to put up its share of the revenues too much.

It was a model that worked well for the first half of the twentieth century, but after the Second World War it began to sour. In the 1950s the Iranians tried to nationalize the company and drive a much harder bargain.

Although, as we saw in Chapter 5, nationalization failed,[5] the balance of power shifted. To survive, BP needed to turn elsewhere. The second BP was built in Alaska and the North Sea. In Alaska, it succeeded after many repeated failures in the late 1960s, and in the end almost by mistake. It was about to give up when its luck turned at Prudhoe Bay, and it found itself with a big new basis for its survival and subsequent growth.[6]

But Alaska, whilst giving a long tail of production, could not keep pace in terms of new reserves. BP's other break came in the late 1960s. As a British company it was drilling in British waters and discovered the West Sole field in 1965.[7] It took the OPEC shocks of the 1970s to turn what was at the time a highly marginal, even experimental, possibility in drilling in shallow waters offshore into its next foundation for its reserves alongside Alaska.

As the 1970s turned into the 1980s, and as the British government sought to milk the North Sea oil and gas production for all it could, BP came under pressure to explain to investors where the next tranche of reserves would come from, post-North Sea. Its answer was the Gulf of Mexico and Russia. The Gulf came its way partly through the great wave of M&As in the 1990s, in response to the collapse of oil prices from 1986. But the Gulf posed big technical problems, and big cost problems too. Much of the reserves were in very deep waters, and this needed new technology and sufficiently high prices to justify the huge sunk costs and the expertise of the oil service companies. Just how risky and technologically demanding was revealed in the 2010 Deepwater Horizon accident. It was to take horizontal precision drilling to intercept the damaged well in 5,000 feet of water, and the cost to BP was eventually over $50 billion in damages and remedial works. The company was nearly broken as a result, and dramatically diminished in size and importance. Yet its size enabled it to (just) withstand a shock that would have bankrupted all but its small number of peers.

BP's second bet was Russia. As we saw in Chapter 6, in the 1990s, after the Soviet Union had collapsed, Russia was a much-reduced economy. Its one major asset, its fossil fuel reserves, was seriously affected by both the post-Soviet chaos and also the (related) low oil price. Yeltsin needed money, and the oil and gas industry needed investment. John Browne saw the opportunity.

It was one of the few frontiers left in the world where the company could actually own the reserves, book them and therefore hold up its reserves-to-production ratio. But it was a frontier not only in technical terms but also politically. BP got out very large dividends in the early years, recouping its investments. But the game changed when Putin came along and the oil price recovered. Now Putin wanted to take back control of the natural resources, and BP was eventually to escape with a minority 20% stake in Rosneft, Russia's new national champion.

The last hurrah and the coming of shale

Finding new reserves to book looked harder and harder in the 2000s. War in Iraq, sanctions in Iran, and religious fervour across the Middle East reduced the opportunities. Africa offered lots of opportunities, and Nigeria and Angola already had significant industries but the oil companies faced both NOCs and the entry of China. They were increasingly forced offshore in the search for new reserves. The East African coast and the Arctic appeared to offer potential, but these were on the outer envelope of costs and technical feasibility.

Just when things looked particularly bleak, along came shale in the US and the tar sands in Canada. With oil at $100 per barrel, the costs looked manageable even for the tar sands, and the US was blessed with its unique property rights giving the surface owner the rights to the subsurface resources. The shale bonanza of both oil and gas became a surprising and significant new opportunity for the companies, and the sheer speed with which the industry developed and the outputs increased were a lucky break, and a technological wonder.

Shale changed the game once again for the companies. On the one hand, the new technologies opened up a cornucopia of possible future reserves. Now there were opportunities across the world, from the US and Canada to Argentina, Australia, Algeria and China, to add to the further increase in potential in the conventional oil- and gas-rich countries in the Middle East and also in Russia. Replacing production with new reserves no longer looked so difficult, especially if the shale companies could be bought up, thereby booking their reserves. On the other hand, shale's economics were quite different, and importantly very different from those that the big companies had structured themselves to handle.

Shale is all about multiple wells and high depletion rates. It is more like slash-and-burn agriculture than the large-scale fixed and sunk costs the oil and gas industry is familiar with. Shale is all about having lots of rigs which can be moved fast. They can be deployed and withdrawn as prices move up and down. Thus, as the oil price halved, the rig count became an obsession for market analysts. How many rigs have been taken out? How many added?

This is a million miles away from a conventional well. Oil price falls will not lead to the temporary mothballing of existing wells. On the contrary, given the need for revenues to cover fixed and sunk costs, lower prices might even induce higher output to hold up the revenues, as discussed in Chapter 1, and as witnessed in mid-2015 across the Middle East.

This very different cost structure enables different kinds of companies to come into the market. Shale was developed by small and middle-sized companies in the US, starting with Devon Energy, the vehicle of the father of the modern shale industry, George Mitchell, using a lot of debt finance rather than balance sheet depth. This was as much a finance story as a technical one.[8] The big companies came in later, and there were impressive takeovers. Shell and Exxon purchased both shale and tar sands assets. Even the Chinese came into Canada, making a spectacular mistake on valuation with the acquisition of Nexus. These companies all paid big premiums, and they all got severely burnt as a result.

There can be little doubt that there is a lot of shale oil and gas around the world, and that the Arctic is rich in conventional deposits too. There is, contrary to the concerns of the peak-oilers, and indeed many of the companies in the pre-shale 1990s, plenty of the black stuff out there. But, just when the potential and opportunities have opened up on the supply side, the companies now confront a very different problem: is it worth doing the E&P? Might the great oil age be over, to be outcompeted first by gas and then by new technologies? Does oil have a future role in its two core markets: transport and petrochemicals? Will demand leach away, just when the supply side has come good? What should they do?

The oil-to-gas substitution: oil companies turn themselves into gas companies

Option one is gas. At first sight, the fact that gas is growing market share, and that oil might lose out to gas, might not seem like a big problem. Oil companies are all in fact oil and gas companies and, as noted, gas developments have much in common with conventional oil. Gas was seen for much of the twentieth century as an inconvenient by-product, in many cases simply to be got rid of by flaring it off. Might not the great oil companies become even more focused on gas?

The gas story works with the three big changes described in Part One. The fall in fossil fuel prices is worse news for oil than for gas; decarbonization places an imperative on getting out of coal, and gas can play the role of a transition fuel for electricity generation; and it will take time for the new technologies to cut into the electricity generation markets.

The obvious strategic lesson is for the oil companies to turn themselves more and more into gas companies. Some have understood this. Shell's acquisition of BG Group, completed in 2016, is aimed at taking advantage of Australia's growth into a dominant player in the LNG market, second only to Qatar.[9] But from an investor perspective this makes sense only if the company doing the switch has a competitive advantage in gas. The alternative is for gas companies to grow and oil companies to shrink: Gazprom against Rosneft, Gaz de France (GdF – now called Engie) against Shell, and so on. Whilst the oil companies might have the core skills for developing conventional gas fields, they may be less good at running pipelines, LNG facilities and LNG tankers and gas storage. Some might be good at shale (gas or oil) with the same sorts of skills, technologies and capital structures. Large oil companies are not the only, or indeed always the obvious, players here.

The oil-to-gas substitution is the strategy not just of Shell, but also of other primarily European oil companies such as Total and Statoil. It makes these companies look better from a carbon perspective if they can target coal and make the carbon argument more one of coal-to-gas in electricity generation. In practice, in Europe, and notably in Germany, the opposite is the case: gas is giving way to coal. Companies like Statoil, keen to supply Europe's CCGT markets, have been constantly disappointed. Gazprom has faced falling overall gas demand from Europe. It is only in the US where the switch is pronounced – and there it is shale gas that is making all the running.

It is in any event only a stay of execution. The carbon constraint means that gas can play only a temporary role, and this temporary role is a race against technical progress in next-generation renewables. There may be a need for peaking services and back-up for the renewables for a long time to come, but there is a difference between installed capacity and the total amount of gas burnt. The electricity systems may need more gas capacity, but not necessarily need more gas.

The total carbon bubble, the stranded assets debate, and divestment campaigns

Gas may be a bridge fuel, and it may bring the big companies some relief as their oil reserves are both harder to book (as the NOCs force them out) and of potentially lower value as and when prices fall – but it is only temporary. In a decarbonized world, gas eventually gets pushed out too. The predictable surprise of the end of the super-cycle has seriously damaged oil company market values and squeezed the marginal new frontiers, pushing the market back to the OPEC core. Despite these companies struggling to cope with such immediate challenges, and the possibility that many of their current investments in deep water and tar sands might remain out of the market for the foreseeable future and perhaps for ever, these market shocks are but temporary compared with the required demise of oil and gas in a decarbonized world.

If decarbonization is to happen, and if 2°C (or even 1.5°C) is to be the maximum temperature increase, it is relatively easy to work out how much more oil, gas and coal can be burnt before breaching the global concentrations of carbon in the atmosphere. As discussed in Chapter 2, this has been calculated, and it is a crude quantity-based argument. It takes no notice of price and ignores the investor perspective on valuation: price multiplied by quantity, discounted by the cost of capital.

A more subtle valuation issue arises long before the carbon constraint bites. It concerns the implications for investors if the price of oil continues to fall into the medium and long terms. The recent price crash illustrates this: the value of oil companies has fallen substantially without any quantity adjustment and no serious change in the cost of capital. Oil in mid-2017 is simply worth about half as much as it was in mid-2014. Investors in oil (and indeed coal and other commodities) have lost a lot of money. It has nothing much to do with carbon and climate change. It is just a change in price that has reduced value. It is the sort of thing that changes asset values all the time: movements in prices cause changes in the value of assets. There is nothing 'stranded' here: it is just how markets work. For oil-dependent countries it can be much worse: Saudi Arabia's national wealth has taken a big tumble.

Where the physical constraint matters (if it is binding) is in two critical areas. The first is that the fossil fuel industries are doomed in the long run. There is not much of a long-term future in this business. It looks more like the high street booksellers' market: a small specialist niche, no longer one of major high street stores. The uncomfortable conclusion is that Exxon, Shell and BP will not be around in anything like their current form by mid-century. The implication is that the reserves-to-production ratio is going to be redundant as a measure of value for investors: beyond the envelope, the reserves will be worthless, because they cannot be burnt.

The only possible escape is to find a way to burn fossil fuels at scale without emitting more carbon. In theory, CCS would do the job. And in theory, we could even simply take carbon out of the atmosphere generally. But in practice this looks like an unlikely saviour. As noted in Chapter 2, the sheer scale of what would be required, the absence of sufficient holes in the ground, and the fact that the carbon and other gases have higher volumes than the carbon in the fuels extracted together make this an unlikely winner with investors. Better from an investor's perspective to simply run down their reserves, pay out the dividends, and eventually close down.

The second critical issue is the speed of the decline for each of the fossil fuels. Here the issues are about the carbon benefits of switching from coal- to gas-fired electricity generation discussed above, and from oil to gas in transportation and petrochemicals. For this to work, the coal has to be forced out, requiring a combination of emissions performance standards being imposed on coal power stations (which force through closure and make new coal-powered stations uneconomic by requiring CCS) and carbon pricing to reflect the relative carbon densities of coal and gas. Government intervention is also required in respect of electric cars. They are in the hands of the politicians.

As the forecasts of the major oil companies show, they don't believe much of this will happen before mid-century at the earliest. Most forecast demand growth for all the fossil fuels through to mid-century. And to the extent that they do believe it will happen, they have begun to lobby against coal (which is an obvious win for themselves) and for a carbon price to help in the relative costs of coal and gas.

One final dimension of the stranded assets argument might explain why the companies are so unconcerned. Whether decarbonization happens or not, it is not going to have much of an impact on the demand for oil in the next decade or so. From a business perspective, long-run demand is a matter for the long run, and at a typical positive discount rate that companies apply to thinking about assets, projects and future reserves, the long run does not much matter. More important to companies and investors are the dividends *now*, and these are more vulnerable to the immediate changes in oil prices, immediate political crises, and whether, for example, China's demand growth is tailing off or will fall off the cliff completely.

The future matters from a societal point of view.[10] There should be some down payment now for the damage done to the climate (and indeed biodiversity). Yet from an investor perspective, why, for instance, sell shares in Shell yielding, say, 6%, for fear of a decline in the demand for oil in one or more decades' time? The reason to sell might be that the oil price could go lower in the next few months, not what happens after 2030.

This is what the divestment campaigners get wrong. The assets of oil companies (their reserves) are stranded only in an undiscounted future. From a narrow investor perspective there is no reason to sell now. But then this is not the main point of the stranded asset campaigners. Rather, their objective is to bear down on what might be called the 'social licence to operate' – to turn oil companies into the bad guys, and to persuade investors that it is unethical to invest in them. This could hurt: the wider public (and their trustees in the big pension and endowment funds) might decide that the ethical dimension should override their narrow concern with returns and subsequently divest.

The comparison is often drawn with the divestment campaign directed against Barclays in respect of its interests in South Africa in the apartheid era. Yet this is not a good example. In the case of oil, the IOCs from which these investment funds can divest are now dwarfed by the NOCs. The NOCs could simply step into the space created by the exit of the IOCs. The withdrawal of Barclays from South Africa would have been a serious blow to the apartheid regime. Oil production globally will not fall because of the stranded assets divestment campaigns against IOCs. How exactly are the divestors going to get out of Saudi Aramco (even if 5% is privatized)?

Worse still, divestment will create opportunities for less 'ethical' investors. If it were to impact on the share prices, it would mean that those who ignored the moral arguments would get a higher yield. The shares would be an even better bet. Take a look at tobacco company shares. Smoking is already killing more people than climate change, and probably will for the rest of this century. Those who choose for ethical reasons not to hold tobacco shares have had no obvious impact on the companies' financial performances, and they have forgone what have been very high-return investments. Better, in both cases, to campaign for tighter regulation and taxation.

A further mistake that the divestment campaigners make is to believe that their ethical behaviours will change boardroom behaviour. They must really believe that companies will stop trying to find more oil and gas for fear of divestment. Whilst there is a place for 'corporate social responsibility', and all companies pay lip service to it, its impact has been pretty minimal. What matters are the fiduciary duties of the directors, the rules of the game that governments lay down, like carbon taxes and emissions performance standards, and what is happening in the markets. Shell may well have stopped Arctic drilling, and the stranded assets and divestment campaigners claim it as vindication of their actions. But the real reason has much more to do with the costs, the nature of the wells drilled, the ending of the commodity super-cycle, OPEC's outputs, shale, and the weakness of demand, including transport and petrochemical demand, than it does with the narratives of 'eco-warriors'. And it has not stopped a host of companies from bidding for licences elsewhere in the Arctic. Oil companies may be on a long-run declining path because the demand for oil will in due course tail off. No amount of E&P in the Arctic will solve these market and technological fundamentals.

The deadly impact of competing technologies

The companies are almost certainly right to believe that it will not be policy that drives out fossil fuel anytime soon. The rules of the game are changing at glacial speed. They are right too to dismiss worries about their assets being stranded once discounting is taken into account. There is little evidence that it is energy policy or investors' worries about stranded assets that are going to make the difference and kill off oil and gas for some time to come, if ever.

On the contrary, around the world, the increase in population, the growth plans of developing countries like India, and China's hitherto somewhat cavalier treatment of the environment (except where urban air pollution is concerned) do not point to a significant change of direction. As noted in Chapter 2, the Paris COP displayed just how vacuous international top-down agreements are in this context. The NOCs will take up any slack left by the 'disinvestors' in the IOCs, and since they are government-owned, the only plausible disinvestors are ultimately political leaders like Putin, King Salman of Saudi Arabia and Xi Jinping. The suggestion that Saudi Aramco might be worth $2 trillion, and the excitement caused by the idea that 5% might be privatized, does not look like a convincing endorsement for the divestment and stranded assets arguments. For the companies this looks like very strong ground.

But in looking to climate change policies, all the oil companies are missing their key challenge. It is unlikely that we will stop burning fossil fuels because governments or ethical investors decide we should. What will change the game is if there are cheaper alternatives. As indicated in Chapter 3, that is exactly what may happen – and sooner than many of the companies realize.

The strategic implications of technical change are poorly understood. Few companies in long-established and stable markets see big changes coming, and few are much good at adapting. The internal cultures typically have a built-in bias towards the status quo, and their leaders tend to focus on this rather than the more distant horizons. In the case of oil and gas, prior to the development of shale technologies, technical change was incremental and within the paradigm of the industry. The oil and gas fields have remained their core focus, and technology has been about getting more out of the existing wells through maintaining pressure, getting better at drilling, and building improved above-ground infrastructure. The move from onshore to shallow and then deep offshore has evolved in a symbiotic dance between the oil price, the incentives for R&D and the results.

The sorts of technology that might displace the final markets for oil – electric cars, next-generation solar and even nuclear, and new materials – are disruptive in the sense that they come from *outside*. Unless there is very fast insider technical change, the company model is radically challenged. Shale technologies show how inside technical change can respond to price, but it cannot do much about the possibilities like opening up the light spectrum and the development of solar film.

In the oil business, a number of companies have tried to develop 'alternative energies'. This is the second option. BP's attempt to go 'beyond petroleum' is the classic example. While, as noted above, John Browne may have recognized the need to move on from the North Sea to Russia to find new reserves, the 2000 'beyond petroleum' exercise was not a success.[11] Its failure was not for lack of effort: money was thrown at the problem, a separate renewables division was set up, and several billion was invested.

What the example reveals is two things. In a company where the main revenue drivers are large-scale oil and gas, alternative energy is never going to have the boardroom priority or attract those who want to succeed in the company. The 'success' areas are likely to be those nearest to the core skills, rather than wholly different technologies. The most effective strategy for getting to the top of a big oil company has been to work for years inside it. Insiders lead most of the main oil and gas companies, and for good reason: they know a lot about the assets and about how things are done. They know the ropes. They know all the key people. From time to time it might make sense to parachute someone in from the outside, but even in these cases they tend to be industry people. Where there are exceptions, for example in NOCs, the outside appointees tend to be political or part of the oligarchic structures. Here the game is about extracting rent, political influence and the sorts of insidious corruption witnessed in many resource-cursed countries. Russia is a good example. Rosneft and Gazprom have been led in this way, with very predictable poor results for the efficiency of the businesses.

For large oil companies, alternative energy is a branch line, unless the directors really think it is going to be the future of the company. The clue is in the name *alternative*. A cursory glance at the revenue lines in BP's accounts tells us that this would have been an enormous ask: alternative energy would have had to grow in the company at a rate of around the total world growth of alternative energy to make much difference to its bottom line. Aspiring managers climbing up the company's greasy pole would know this. The best hope for them personally would probably be if the business were floated. The implication is obvious: 'beyond petroleum' was never going to attract the future leaders of BP. The latest oil company to try this line is Total, promising to devote 20% of its business to renewables by 2035. That leaves 80% in fossil fuels, where the cash flows and dividends will be concentrated, after another two decades have passed.

To the extent that alternative energy could have been within BP's and other companies' grasp, the answer lay with biofuels. The reason is largely negative: the companies are legally obliged to blend biofuels with conventional petrol and diesel for transport. More positively, biofuels need refining, and BP and other oil companies have refining skills. They have to do it, and they have the core skills to do so. Yet even here, the processing of sugar cane, corn, wheat and rapeseed, and other food crops is as much an agricultural business as an oil company one, and the difference between diffuse supplies of bulky materials and the economics of large oil fields is very considerable.

The implication is clear: whilst an oil company might turn itself into a gas company, there is little prospect of it turning itself into a solar or nuclear business. It just does not have the core competences, and indeed the core competences it has probably obstruct its ability to make this kind of switch. There is little the IOCs and the NOCs can do about disruptive energy technologies, and their shareholders will probably prefer that they stick to their knitting. The fact that as a result they might slowly wither away over the next decades is not an investor problem: investors can gradually switch too, benefiting from the dividends as the companies decline, rather than having profits diverted ineffectively into alternative energies. The investors can put their money elsewhere, with new companies with the relevant core competences in the new markets and technologies, and with their more targeted management. If the oil and gas companies are forced to gear up further to prop up the dividends, so much the better, since they will have less scope in the future to divert cash flows into renewables and other diversifications.

What if oil companies believe that they are doomed in the long term?

If the oil price is going to stay low as a result of the ending of the commodity super-cycle, if decarbonization is going to happen in the medium term, and if new technologies are going to be cheaper than oil by mid-century, what should an oil company do?[12]

Options one and two are of limited appeal: the first, gas, gives only temporary relief, and the second is likely to fail. Option three – harvest and exit – has some similarities to that confronting the oil-producing countries discussed in Part Two. Whereas in the past the expectation of ever-rising prices meant that increasingly costly E&P would come good, this no longer works. Nor does holding back production now on the assumption that the value of the reserves will keep rising with the higher prices: the value of reserves would fall over time, particularly if a discount rate is applied.

The strategic implications are radical. Right now, many companies are trying to hold on to the key E&P projects on the assumption that prices will eventually rise. They plan to see out what they regard as a *temporary* downturn. Chief executives keep referring to current conditions as challenging, as we saw in Chapter 1. They see themselves as holding their nerve in adverse market conditions, and they want investors to stay with them. But if oil prices are assumed to gradually fall over time as demand slowly withers away, to be overtaken by other types of energy, then the high-cost frontier E&P projects in areas like the Arctic should be scrapped – permanently – and the companies should wind down those parts of the businesses. Under option three, the management focus shifts to getting the maximum out of existing resources. Put another way, to carry on with the frontier E&P is to believe that decarbonization will not happen on a significant scale and that disruptive technologies are not going to work.

Option three also implies a response to a profile of falling future prices (that is, oil produced today being worth more than oil produced tomorrow). It is to maximize short-term production. This strategy is all the more attractive if countries read the same message and also decide to maximize production. If, as discussed in Part Two, key Middle Eastern countries take this view, the supply from very low-marginal-cost oil fields will go up. This extra supply on the country as well as company side will in turn drive down the price, making the outcome self-fulfilling. The Arctic would look even less appealing.

In response to the price falls at the end of 2014, oil companies opted for cost-cutting, and no doubt they will reap considerable efficiencies as a result. But the above considerations indicate *where* they should be cutting costs. This also radically alters the way oil companies are valued. Instead of concentrating on the reserves-to-production ratios discussed above, and assuming that the existing reserves will go up in value, they should concentrate on valuing a gradually declining profile of returns against a scenario of gradually falling long-run prices. It should be all about dividends. Ultimately, the best strategy might be option three – harvest and exit. Few corporate boards, however, are willing to preside over a declining business, so such a strategy typically gets imposed from outside as dividends come under pressure from lower revenues – and after the companies have exhausted all the alternatives. The responses so far, including borrowing to pay the dividends, suggest that it will be no different this time. Slow death is the most likely outcome, possibly dragged out over several decades.

CHAPTER 10

Energy utilities
A BROKEN MODEL

If the future is increasingly electric, it might seem obvious that the incumbent electric utilities will be the big winners. But this turns out to be unlikely. Today's big electric utilities are largely fossil fuel-based, and their vertically integrated structures reflect the fact that there is little storage and the demand side is largely passive. As described in Chapter 3, all this is likely to change as a result of decarbonization and new technologies.

The impacts are already becoming apparent. The two biggest German utilities are in big trouble. RWE and E.ON have more than halved in value in the last few years, and it would have been even worse for RWE had its lignite mines and coal-fired power stations not enjoyed a brief Indian summer. For both, the battle to maintain their large integrated structure has been lost and they are breaking themselves in two. France's EDF has been hit hard by falling wholesale prices as well as by the specific nuclear issues. Its market value fell from over €100 billion to less than €20 billion in just a few years. The scale of value loss by European electric utilities is extraordinary. The integrated companies struggle to find a way of coming to terms with the renewables spilling onto the systems they once controlled, and try to find ways of hanging on to their customer bases.

The integrated energy utility model is nearly as old as the oil companies, though it started in a very different place – largely locally through municipal public ownership. In Europe this model remained publicly owned for most of the twentieth century, whilst in the US, rate of return regulation created a system of private ownership under public control through commissions and regulators. Developing countries have largely followed the regulated public ownership and/or control models too.

Why did the energy utilities evolve as they did? What does zero marginal cost renewables mean for their grip on electricity generation? What will storage do to them? How will they cope with decentralized generation? How will they compete for households, with the rollout of smart technologies in buildings and homes? Can they survive?

Why the vertically integrated large utility model worked: the special problems with electricity

The core economic problem that electric utilities share with oil and gas companies is that they both have large sunk upstream costs, with long asset lives. They are similarly vulnerable to price volatility, and in particular to marginal cost pricing, since their average costs are higher than their marginal costs.

The common solution to this problem has been to try to fix the market, and thereby to manage the prices. As we saw in the previous chapter, for the oil companies this meant collusion upstream, control of the refineries and pipelines, and control of the retail market. They built themselves around the full vertical chain, captured the customers, and therefore were able to pass through their costs – hence recovering their sunk and fixed investments. The result was that short-term spot competition was kept at bay for most of the twentieth century, and the oil companies could plan out for the medium-to-longer term. Unsurprisingly, many ended up looking rather like government departments, using similar planning approaches. BP and Shell, for example, fitted into the British Whitehall model.

In the twentieth century, the electricity utilities faced these problems along with two additional ones: the absence of storage and passive demand. Unlike oil, electricity could not be easily stored, so demand and supply had to be instantaneously matched. Since the demand fluctuated over the day and night, it meant that power stations had to be built to meet the total peak demand. The trouble is that the peak cannot be exactly predicted in advance, and in any event there were bound to be plant failures and unanticipated outages. Hence, to secure the supply to customers there needed to be an excess margin of capacity relative to expected demand, just in case.

Oil markets do not face these problems: oil can be easily stored, from keeping it in the oil well, to storage at the refineries, tanks and tankers, and through to the fuel in vehicle tanks.

The problems that an excess margin poses for electricity companies are twofold: first, it is unlikely to be profitable at 'normal' prices; and second, the overhang of extra capacity depresses the market price. The reason the excess is there is insurance: the insurance that if demand spikes then there will be enough supply to meet this. How should this insurance be paid for?

One route is to just let the prices spike from time to time. At points of stress, the price shoots up, and the supplier of this peaking capacity margin wins the jackpot. For the rest of time, the peaking plant is not needed and hence it earns nothing. It is easy to see that this is not very efficient, and it is telling that it has been relied upon only when there is an inherent capacity margin already on the system, and so the problem does not arise. As soon as it is a problem, and hence prices might spike, the answer is to pay an insurance premium.

Since security of supply is a property of the system and not each individual's demand, the premium needs to be a system one, recovered across the entire customer base, regardless of the individual consumption. It is a bit like nuclear deterrents: a country either has it or it does not. It is a countrywide cost.

The way to pay for this is through a capacity payment, charged to all. But getting everyone to pay requires a way of forcing every customer to do so. Short of general taxation, the answer is that there must be some sort of a monopoly charging base. In the twentieth century, this monopoly was often statutory. In Britain and France, for example, it was illegal to compete against the electric utilities.

The second major difference with the oil companies is on the demand side. In electricity, customers are largely passive. They cannot easily vary their demands in the short term, and there are few alternative ways of lighting buildings and houses and driving electric machines and appliances. Therefore, rather than the customers being able to respond to prices, they relied on the electric utilities to do this management for them. In effect, customers got an on-demand long-term contract with a stable price, which was delivered through the electric utilities' investments in generation, transmission and distribution.

With a monopoly there are lots of possibilities

The monopolies had a number of advantages that they could exploit. On the one hand, they could plan. They could take a view of the electricity systems *as a whole*, integrating decisions about power station investments and management with the main transmission grids and the distribution networks. They could and did go even further upstream. The monopolies could not only decide what sort of power stations to build and where, but they could also contract with the fuel suppliers on the basis of long-term contracts, which could in turn underwrite the sunk and fixed costs of the oil, gas and coal companies, and later the nuclear fuels and technology companies. This allowed the sharing of risk and therefore benefiting from stable prices for fuels inputs, which could in turn give passive customers further price stability.

In the British example, the Central Electricity Generating Board (CEGB) contracted with the National Coal Board, and the National Coal Board planned its coal output on the basis of a certain demand and revenue. The CEGB had a monopoly on generation, and the National Coal Board had a (imperfect) monopoly on coal mining. Later, as a statutory monopoly British Gas could sign long-term contracts with North Sea gas companies and manage the development of North Sea gas against the growth of its transmission and distribution system and the growing customer base. All gas landed from the North Sea had to be sold to British Gas.[1]

With the ability to plan, the companies developed the tools to do so. They had planning departments, and these got linked in with economic plans for the country as a whole. It was the sort of structure now familiar in China: a five-year high-level plan linked to sub-plans for the number and type of power stations to be built by China's state-owned electricity-generating companies.

The vertical integration required was complete: from the captured customers through the network systems, to the generators, and then through long-term contracts (which are themselves a form of vertical integration) into coal, upstream gas and nuclear.

So far, so good. But there was a big downside to this model. Monopolies could exploit their customers, using their market power to extract excess profits and to price-discriminate. They could also opt for what economist John Hicks called 'a quiet life'.[2] They did not need to be particularly efficient and could spend on their management and their perks and interests. It was a structure ripe for capture by vested interests: in the British case, by the management, the trade unions (especially the National Union of Mineworkers) and the nuclear industrial interests;[3] in the case of the current Chinese state-owned companies, by the party officials.

In the absence of competition, the answer to the profits problem is regulation, and for most of the twentieth century this included an allowed rate of return – a system that still survives explicitly in the US, and explicitly or implicitly for network regulation across the world. All sorts of fancy formulae have been developed to incentivize the monopoly parts of the businesses, but they ultimately all end up limiting the rates of return, some *ex post* and some *ex ante*. Even price-cap regulation, pioneered in Britain, is really rate of return regulation with a lag.[4]

Price discrimination is rather different: if the electricity utility is building out its customer base, it makes sense to charge closer to marginal cost at the periphery for marginal additional customers, and more to the captured core customers, provided that extra customers benefit all by making some contribution to the fixed costs. In the economics jargon, Ramsey pricing is efficient: prices should be inverse to the demand elasticity.[5] This turns out to be relevant again for two reasons: there will be a much greater emphasis on fixed system costs in the future; and there are new customer bases to build, notably for electric cars and for decentralized electricity systems.

For most of the twentieth century the electricity industry was dominated by these large vertically integrated monopolies that were often owned by governments and always regulated. Despite two key breaches in the model it survives today: big vertically integrated electric utilities continue to dominate across the world. The breaches were the removal of the monopoly on generation; and the liberalization of the retail market to enable customers to switch supplier. Both have had an impact, but are yet to make a really radical change to the models of the past. But they will, as we shall see.

Zero marginal costs and fixed-price contracts

The efficient scheduling of power stations should be done on the basis of short-run marginal costs (SRMCs) – the extra cost of adding each extra unit of output onto the system, given the capital stock of the power stations.

These extra costs are mainly the costs of the fuels (since the capital cost of the power station itself is fixed). As demand rises during the day, more power stations need to come onto the system and generate, and these are brought on in the merit order of their SRMCs, starting with the cheapest and progressing up the supply curve. The price is set by the last station needed to meet the demand at each point in time: this is the system marginal price (SMP). The companies get paid this SMP, which is equal to or above the SRMCs of all the stations needed at any point in time. The difference between their own SRMC and the SMP contributes to their fixed capital costs.[6]

SMP provides revenues to the generators, but it is not enough to cover the full system costs of the excess capacity needed to ensure security of supply, and hence the companies also receive a capacity payment on top.[7] In the twentieth century, almost all electric utilities received an energy (SMP) revenue and an additional capacity revenue, and customer charges in turn reflected these (plus the transmission and distribution costs). A wide variety of specific tariffs achieved this outcome across the world's electricity industries. In Britain, it was via a bulk supply tariff charged by the CEGB, and passed through to customers paying fixed 'standing charges' and variable 'energy charges'. The result was very different from paying a single set price for fuel at the filling station, and hence highlights a strong differentiation between the economics of oil and gas and the economics of electricity.

This twentieth-century system relies on the fact that there actually is an energy cost, and hence an SRMC on which to construct a merit order and derive an SMP. But what if electricity generation has zero marginal costs? If SRMC is zero and hence SMP is zero, then the wholesale price would turn out to be zero. The whole edifice collapses. How then would generators get paid? This is the prospect that the companies now face, and its implications are very radical.

Nuclear and most renewables (but not biomass) have this characteristic. When the wind blows and the sun shines the energy bit is free. There is no fuel charge for solar radiation. As long as there is a bit of non-zero marginal cost generation still needed at all times, it might not much matter. But once there is enough zero marginal cost generation to meet the *total* demand from time to time, the wholesale price will be zero from time to time too. This is already beginning to happen in Germany with particular configurations of good wind flows, lots of sunshine and low demand.

As decarbonization gets going, this outcome will be more frequent and eventually happen most of the time.

All sorts of problems arise in this new model for the conventional vertically integrated utilities. Even when the zero marginal cost generation does not meet total demand, it still meets some. This means that more expensive plants at the top of the merit order are not needed. The SMP is lower, meaning less contribution to the fixed capital costs. It pushes some power stations off the system altogether, mothballing or closing them earlier than anticipated. This has already been devastating for E.ON and RWE. In effect, the renewables cannibalize the fossil fuel plant, and the lower prices that result expropriate the electric utilities' shareholders. They have to write off their assets. This is indeed what has happened, big time. Both companies are worth just a fraction of what they were a few years ago, and their dividends are being undermined as a result. They are not alone. For EDF it has been even worse. All the big players have suffered as a result, and the share prices reflect this. The wholesale price has fallen dramatically – the exact opposite of what so many politicians and company executives so readily assumed up until very recently.

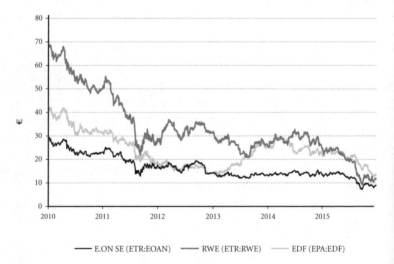

Figure 10.1 Major European utility share prices 2010–15

Sources: Yahoo Finance; Google Finance

It no longer makes sense for the electric utilities to plan and build new fossil fuel plants across much of Europe, and the existing ones earn less as the wholesale price falls with the increase in renewables. This will spill out across the world's electricity markets in due course. Europe has led the way in this area of climate change policies, and has provided a textbook example for others to note of the consequences of its particular path. The existing fossil fuel assets get stranded. So the conventional electric utilities have to decide how to build their future. Unless someone pays them to have the excess plants on the system to meet unexpected demand, this is a declining business. E.ON has recognized this by putting all these fossil fuel assets into a separate company – Uniper – which has been somewhat inaccurately compared to a 'bad bank'. Shareholders can take a declining dividend on these assets, as this part of their business slowly dies.

There remains a business in *peaking* plant, and though it is sometimes conflated with the excess capacity problem, it is in fact a very specialized niche. Peaking is all about the intermittency of wind and solar. It is not unpredictable in general, but it is at particular times, until wind and cloud cover can be forecast precisely. The power stations required to solve this sort of problem are typically peakers, designed specifically for this purpose. They are not the conventional gas and coal plants. They compete with storage options and with demand-side response mechanisms. This is not an area in which the big utilities have particular expertise, and it tends to be populated with small, focused niche players.

Faced with the gradual death of the conventional big power station business, as zero marginal cost gradually takes over, an option for the big electric utilities is to follow the maxim 'if you can't beat them, join them', and switch to the new model. A bit like BP's attempt to go 'beyond petroleum' described in the previous chapter, the electric utilities could go 'beyond fossil fuels'. But, like the large oil companies, they suffer from similar disadvantages. Nuclear is one option that the big electricity utilities might like to develop (and both RWE and E.ON originally did try),[8] but there is little prospect of much new large-scale nuclear in Europe or the US. China, India and the Middle East are where nuclear might flourish. Building small-scale onshore and offshore wind and fitting solar panels are hardly skills that companies specializing in large-scale power stations are likely to have. Only a few have managed this transition – notably Spain's Iberdrola with its Scottish Power subsidiary, and the Danish company DONG.

Even in financing the renewables, the big electric utilities now face competition head on with infrastructure funds and other financial institutions. The revenue streams from current renewables are derived largely from state-backed feed-in tariffs (FiTs), and they lend themselves to securitization once the projects have been built. This is a business more for pension and infrastructure funds looking for proxies for government bonds, rather than those more equity-driven upstream power station businesses and their wholesale market revenue streams.

In this new generation market, which will be described in greater detail in the next chapter, vertically integrated companies have few obvious competitive advantages, though they are trying, with mixed success. Bidding for fixed-priced renewables contracts is open to all, as is the market for peaking capacity. As with the oil and gas companies, 'harvest and exit' may be a better option for their shareholders, paying out cash as dividends and minimizing capital expenditure. Investors can invest separately in the more specialist renewables energy companies, and optimize their portfolios and risk management accordingly. They do not need the electric utilities to do this for them.

System operators and the grids

The zero marginal cost world, which is the consequence of the technologies developed to meet the decarbonization objective, not only undermines the electric utilities revenues but creates other new challenges too. The problem of coordinating the system, which the vertically integrated utilities managed in the twentieth century, does not go away. If anything it becomes more acute, given the impact of the intermittency on the system. This is a function of the system operator and not the integrated electric utilities, which once bundled all these functions inside their vertically integrated monopolies.

As the companies have been gradually unbundled, this system operator function has been typically assigned to the transmission grid, which has to manage demand and supply and make sure the system always balances. Where once the system operator was a function that required control of the generation plants to schedule them in a merit order, the IT emerging in the 1980s and 1990s meant that this could be done by bidding and markets. What had once been SRMC – i.e. costs – now could be treated as prices.

Scheduling plant and ensuring that short-run demand and supply are balanced should get easier as more options for managing the peaks emerge from the new technologies. If electricity can be stored (more on this below) the electricity market may look more like oil and gas. If the demand side becomes activated through the sorts of smart technologies discussed in Chapter 3, this can be flexed instead of bringing more capacity onto the system to meet peak demand. The electricity market gradually loses its 'special' features, and the need for the vertical structures that go with them.

None of these developments alters the basic fact that security is a *system* property, and hence there have to be system decisions, and someone has to be in charge. The bit that survives from the old electric utility model is the central planning function of the system (the central buyer), and hence the role of the system operator. The problem for the existing electric utilities is that the system operator function requires few if any assets. Its command-and-control management of existing and future capacity can be driven by contracts and auctions and prices. It turns out that the actual business of the system operator is quite small, with low revenues. It can be public or private sector (or a mixture of both) and it does not have to be carried out by the transmission owner and operator.

Storage

Storage undermines the existing electric model yet further, flattening the peaks in demand and hence flattening the wholesale prices and reducing the revenue to the fossil fuel generators from SMP. In time it may encourage defections from the grid towards self-sufficient own-generation in households and in local communities. The contribution to and of the grid is thereby reduced. Indeed some of this is already happening, requiring regulators to respond to protect the main transmission income.[9]

As described in Chapter 3, storage comes in various shapes and sizes. There is direct storage of electricity on a large scale as part of generation. Pumped storage is an example: when demand is low, water is pumped up hill; and when demand is high the water is released to generate electricity. Hydro schemes can play this role too, releasing water from dams when there is high demand. Even tidal lagoons can release water between tides. Heat storage and compressed air work as ways of storing energy which can be turned into electricity. A small-scale example in the domestic context is night storage heaters. Special heat-retaining bricks are heated up at night when demand is low, and they slowly release their heat during the day.

The more radical forms of storage are batteries. Batteries are the universal way of achieving small-scale storage. They are in torches, laptops, mobiles and in cars. None of these devices can work without them. They are, however, small-scale, and have little or no role in managing the system electricity demands. With some tariff structures they might be charged at night, but this impact will be trivial until smart technology and real-time pricing does all this automatically.

As again described in Chapter 3, batteries will make a big difference in cars and transport. Electric cars not only cut into the demand for oil, but they also create a potentially very large-scale storage option. They can be charged up at night when demand is low – a bit like pumped storage and heat storage, but on a massive scale. Cars are a major form of fuel storage now because their tanks are, on average, about half-full. The potential for storage from a system of electric cars could be considerable. The result would be to squash the peaks in electricity wholesale markets, putting another nail in the coffin of wholesale-market-driven electricity companies.

Batteries are not, however, just for cars. They could be fitted into houses (as with the Tesla example in Chapter 3). They could also be fitted to wind turbines and solar panels, turning intermittent electricity generation into permanent supplies.

Combining storage and smart demand-management technologies with small-scale peakers transforms the structure of costs in the industry and the ways in which the revenues flow. The business model is radically different from that of the twentieth century. Indeed, the underlying cost structures are turned on their heads. Instead of very large-scale baseload power stations, there are lots of decentralized smaller ones. Instead of no storage and a significant gap between SRMC and SMP, the gap is closed up – and closed further still by active demand management. The incumbents face the prospect of a flattening of the supply curve and the loss of SMP revenues from volatility *and* the prospect of zero marginal cost, driving the system SRMC and SMP towards zero.

Future winners

The conventional electric utility model is gradually going bust. New models and probably new companies will fill the market requirements. Who will they be? The most likely scenario is is of a number of different types of entrant muscling in on the electricity market, and for a period of crowded disequilibrium with the declining incumbents before the marketplace settles down. It will probably take a decade or more.

The classes of entrant will include: new generators, increasingly competing for fixed-priced FiTs and capacity contracts offered by the central buyer; car companies, entering the electric vehicles market; storage companies, bidding for contracts and building battery market share; new suppliers, offering a range of household services; broadband companies, linking up with smart meter technologies; existing retailers, building broadband-hub-related services; and a host of start-up companies.

At the generation end, the zero marginal cost world is one in which fixed-priced contracts are likely to be the driving force behind new entry. As wholesale costs fall, the already vanishing possibility of a merchant power station banking on revenues from the wholesale market being built and financed will disappear altogether. Almost all new power stations will come to rely on fixed-priced contracts, mediated by a system operator, and backed by government as the central buyer. In Britain, they already do.

The game will be about coming up with technologies and projects that tickle the central buyer's fancy. This might be achieved in one of two ways: the project might win in an auction of contracts, or the central buyer might be persuaded to favour it in awarding a non-competed contract. For very big projects, like the nuclear plant Hinkley Point in Britain, it is hard to work out what an auction could look like. For smaller projects, like wind farms, the key for them is to get a reserved market for their particular technology. As we shall see in the next chapter, there are powerful arguments for having technology-neutral auctions across the market as a whole. However, the reality will probably be that the central buyers will keep on trying to pick 'winners'. For the new entrants, it will therefore largely be a matter of lobbying as effectively as possible. Finding projects in key marginal political constituencies, claiming job creation and getting wider political support will all be part of the game. This might extend to whole political parties, such as the Greens in Germany picking their favourite 'winners'.

Some of the new generation projects will be genuinely 'work-in-progress', in the developmental stage of a new technology, and therefore having a demonstration effect benefit. For many start-ups in generation, this is likely to be a significant route to market, and it will be dependent on governmental support. CCS projects fall into this category, as do tidal lagoons and next-generation nuclear.

The generation market is already replete with new entrants and new companies. A host of wind farms and solar companies have been set up. Companies involved in manufacturing the equipment are moving downstream to test markets whilst trying to find a new demarcation line. In many such cases the manufacturers guarantee the equipment and hence in effect become an integral part of the project itself, since they carry much of the equity risk through their maintenance and support contracts.

As noted, entrants in this space include infrastructure funds and new infrastructure companies. Once a fixed-priced contract has been awarded, the equity risk in the price is transferred to governments (as guarantors) and the customers who are forced to pay the price. With maintenance contracts and guarantees assigned to manufacturers, performance risk is transferred to these suppliers, and with it another chunk of equity risk. Once the developer has attained the land and grid connections and has supervised the project construction, and the generation capacity comes on-stream, there is little equity risk left. The project now becomes much like a regulated utility, and with so little equity risk it can be overwhelmingly debt-financed. It can then be refinanced and sold on.

These new post-construction owners come in various shapes and sizes. There are quoted companies. There are mainstream pension funds and there are infrastructure funds, typically servicing the pension fund market. It is a specialist business, as the investor needs to sort out what risks remain, and which of these are equity risks and what risks have been transferred. For example, if a wind farm receives the fixed price only if it generates, and if there is lots of zero marginal cost generation on the system, it may have to bid negatively in the wholesale market to get on the system, or even be constrained off. If the fixed price is contingent on the wholesale price, then there is wholesale price risk, and more zero marginal cost generation can cannibalize the project. Finally, there is political and regulatory risk. Governments often prove less than credible in their commitments and usually have multiple interfaces with companies. They can therefore change the rules of the game without having to directly and retrospectively renege on contracts, though they do quite a lot of the latter, at least partially.[10]

The range of potential entrants in the storage markets is considerable. There are conventional storage options, like salt caverns for gas storage and depleted gas fields, and pumped storage schemes. Many of these projects are about specific assets and rely on a market for storage to be developed. But, as noted above, the really interesting areas for new entrants are batteries. Car companies have an obvious interest in the development of electric vehicles and a large-scale storage infrastructure. In the case of the internal combustion engine, car companies have left the market for its energy to the oil companies, and the interface has been noticeable by its sharpness. Oil companies and car companies have been thought of as very different businesses.

When it comes to car batteries, the car companies have a deeper role in developing the technology, though battery production has so far not been a core competence in the way that building internal combustion engines has. It is yet to be seen whether the car giants like General Motors, Chrysler, BMW, Toyota, Honda and Volkswagen cross this line. Unlike the electric utilities, many of these car companies have invested heavily in robotics too.[11] Electricity represents an alternative fuel to carmakers, but cars do not offer an alternative market for electric utilities. The carmakers are going with the grain of technical progress whilst most electric utilities have stuck to what they know.

The major car companies may themselves be faced with new entrants. New electric car manufacturers may crowd into their space. Tesla is not only building cars but also marketing batteries directly to households. The business model could include these various other applications of new battery technologies. Next come the battery-makers, and here there are the incumbents but also a host of new entrants and start-ups. Companies like Panasonic, AESC and LG Chem are not well known in the electricity industry. And new companies may emerge as new types of battery are invented. With driverless cars, it is the IT that dominates, with Google and Apple both moving into this space.

Storing electricity is a whole new world, with all sorts of interfaces, of which cars are only one part. Household service companies could offer home storage, solar installers could offer integrated storage options, and this adds another blurring of the industry boundaries, alongside that of cars and electricity. The transformation of demand from passive to active, and the integration of electricity demand with smart informational technologies, makes energy supply in part a data problem and opportunity. Enter the IT companies offering the IT kit: the smart meters and intelligent white goods technologies. Enter the household service companies offering bundled service packages. Enter the energy-efficiency installers and energy-demand managers.

Once energy is seen as a component of the broadband hub in the house or building, it becomes of interest to a host of retail companies. The data itself is very valuable, and it becomes the way to access household demand more generally. An intelligent retailer can now sell extremely targeted services and optimize the householder's consumption of them. It is much more valuable than simply knowing what they look for online.

This new cornucopia of data about consumer behaviour opens up endless possibilities to those companies that can tailor their services, and also those that can spread their fixed costs across a broader range of products. In the former category, there are lots of small niche companies and start-ups; in the latter, there are the giants like Amazon, the mobile and broadband firms and the TV and entertainment companies. Then there are surprising entrants. For example, the data from your smart meter will help delivery companies know when you are most likely to be at home.

What should the incumbents do?

The picture painted of a much more disaggregated marketplace with a crowded and overlapping set of companies, the convergence of transport and IT with electricity, and with all sorts of as yet undeveloped ideas and products, implies the break-up of the incumbent vertically integrated model. This might happen by deliberate policy and regulation; it might happen because the companies choose to do it to themselves; or it might happen just because the entrants take the market segments away from the companies.

The end of the vertically integrated model is something the vertically integrated companies have tried to obstruct and prevent for the last quarter-century or more. They have already lost some of their limbs and been battered by regulators, but so far they have survived. The big players of the 1990s are still by and large on the pitch. The first big challenge came with the forced separation of grids and networks from generation and supply. The main transmission system had rested with the monopoly and oligopoly generators as a core instrument of planning. It was developed simultaneously and in a coordinated way with the decisions about the power stations and their locations. And it was reinforced by the trend towards ever-larger power stations.

The second OPEC oil shock in 1979, and the recession that followed, led to a sharp change in electricity demand as the composition of developed economies began to shift away from energy-intensive industries and the high energy prices encouraged energy efficiency investments. The competition agenda which swept the US, the UK and eventually European markets after 1980 coincided with general excess electricity generation capacity. Competition meant that competitors had to have market access, and that in turn meant prising away the electricity grids and gas transmission pipelines – the motorways that got the goods to market.

Whilst the companies often resisted the pressure to sell their transmission and distribution assets, regulation neutralized them, so the vertical model became just generation and supply even if they still owned their networks. The second big challenge knocked away another plank of the vertical architecture. Liberalization allowed customers to switch, and hence the basis of the long-term contracts underpinning the upstream sunk and fixed costs of big power stations was removed.

The vertically integrated incumbents fought back in two defensive ways. First, they engaged in an orgy of mergers. This was most exaggerated in the British market, where the industry was forcibly broken apart with privatization in 1990. Almost all of the British electric utilities were taken over by their European peers, with the exception of Centrica and Scottish & Southern Energy (SSE). RWE bought National Power, E.ON bought PowerGen, EDF bought British Nuclear (with Centrica), and Iberdrola bought Scottish Power. Across Europe, as we have seen, the big companies entered each other's markets as part of a massive consolidation. Across the US the entire corporate landscape has been transformed. It is hard with hindsight to see this wave of mergers as anything but massively value-destructive for shareholders.

Second, the vertically integrated companies created hedges between their power stations and their customers, and used the new wholesale markets as a vehicle to embed their advantages. Without open pool-type markets (more on this in the next chapter) they managed to all but eliminate the possibility of merchant entry in generation.

The companies were remarkably successful in these defensive moves – arguably too successful, since this negated the need to address the shifting fundamentals and changing cost structures. Having limited the competition, the incumbents now face a brave new world in which the vertical model no longer matches the cost structures. All the main planks that supported it have fallen away.

The choices they now face differ across segments of the vertical chains they once dominated. Companies could choose to be generators only. There are already some models of this strategy. DRAX became a single power station generator-only company by default as a result of a forced divestment from National Power. It has taken on a supply business in a limited way to get access to markets in the presence of the dominance of the big players. GDF Suez (now Engie) acquired International Power, demerged from National Power in 2000.

These are examples of generator-only companies with *existing* power stations. A generator-only model for the future involves investing in the sorts of new generation technology coming onto the systems. This includes at one end new nuclear, and at the other small-scale peaking plants, wind farms and solar businesses. Size matters in the case of nuclear, and the dominant players are the nuclear plant manufacturers, with utility businesses taking on the project front-ends, which are typically new nuclear company vehicles. EDF is the most integrated version, alongside the two big Chinese nuclear-only companies, the China National Nuclear Corporation and the China General Nuclear Power Group. It remains to be seen whether the rest of the EDF vertical chain, including coal power stations and a customer supplier business, continue to make sensible bedfellows.

For the small-scale and zero marginal cost generators, it is not clear what, if any, benefits come from scale. Indeed, scale might be a disadvantage, coming with overheads and bureaucracy, investment committees and the usual large-scale corporate baggage. Few, if any, of the big utilities has proved to have a significant advantage in the renewables space, and indeed their general decline in value has meant that they do not even have any obvious advantage in the cost of capital and the provision of finance any more. The more successful companies have been smaller energy businesses, like the Scandinavian companies DONG and Statkraft. Interestingly, both have had a significant element of public ownership to support their growth, and a strong domestic market.

E.ON is perhaps the most interesting example of a big vertically integrated company that has realized that the existing model is bust, and (as noted above) has chosen to structurally address its strategic problems. It is breaking itself into two, with a new renewables and energy services-driven company splitting off from the old coal and gas generation assets. Yet even here the structural separation is as much through force of circumstance – finding a place to put the various things E.ON currently does – as a radical rethink. Why, for example, do renewables and customer-facing services go together? If renewables have fixed-priced contracts, why are there any synergies with selling households electricity and energy services? E.ON's new structure is designed to deal with what it already has, and is less to do with the new market opportunities. Entrants without the baggage of the past vertically integrated company structures do not have such burdens.

A second example is Centrica. With its share price languishing, and having cut its dividends, it announced a new strategy in July 2015: it would focus on its supply businesses, distributed generation and developing its customer services activities.[12] It would sell some upstream assets in the process. This made sense, in that most of the money comes from the downstream, and the end of the commodity super-cycle left its North Sea assets out of the market. Yet Centrica is also an example of how difficult it is for the vertical players to build customer services businesses. It had previously pushed itself into the energy efficiency market and into smart meters as a way of capturing and tying in existing customers. This was not a great success. Selling boilers might be closer to its core skills, but even here the results have been mixed.

One of the reasons that customer service businesses are so difficult for incumbents is that they are already very big companies. For a small start-up, a thousand boilers is quite a decent-sized business, but for Centrica to have enough revenue to underpin its dividends it needs a much bigger market share. It is not obvious that, as the technologies fragment and services become more individualized on the basis of the mass of data becoming available, the broad and inevitably standardized service model will work. Supermarkets have found this challenging as their rivals have developed key niches, whilst the Amazon model has the advantage of being built from scratch from the bottom up for a web-based consumer market. In all these markets, it is now access to software, IT systems and data management services which gives competitive advantage. None of the vertically integrated companies looks well placed in these technologies. They have no history, and no leaders from the IT sectors.

There remains one obvious strategy, albeit one that few company boards are likely to follow. The big vertically integrated companies could gradually close down, exiting their markets and harvesting revenues from their slowly diminishing markets as they go, and keep their capital investments to a minimum. They could focus on paying out as much cash as possible to shareholders and eventually shut up shop. They could do what the oil and gas companies could also do, as argued in the previous chapter, and for very similar reasons.

There have already been opportunities to follow this strategy. DRAX as a very large coal power station could have decided not to invest in biomass, but simply keep going until it closed, with only maintenance expenditure to keep the plant in operating shape for as long as its market lasted. This may have proved much better from a shareholder perspective. In its defence, it has used the original site and built the new boilers around the existing infrastructure, and is now trying its hand at energy services. Yet these synergies with the site pale into insignificance against the supply chain it has had to build. It is less the conversion of DRAX, and more an attempt to build a new company.

A variant of the harvest-and-exit strategy is to reduce the company scope by selling off certain bits to refocus on particular parts of the vertical chain. Centrica is proactively starting this process; EDF is spinning off assets to pay for its nuclear upstream investments; and E.ON and even RWE are slimming down too. In all cases, however, this is less about strategic choices, and more about being forced into more desperate measures to stave off further dividend cuts, decline and even bankruptcy. Most are still wedded to the 'all of the above' strategy, trying to carry on in all segments of the market.

The residual rationale is that the vertical integrated models spread risks, but these are often financial and investors can do this for themselves. If investors want renewables risk, they can buy into renewables assets, if necessary through funds. If they want generation risk, they can buy generators, or speculate on the wholesale price. If they want retail exposure, there are new supplier-only businesses. The plethora of opportunities for investors allows the fragmented parts to be put together in much better financial portfolios than these companies can provide with their particular assets, and which are a product of their particular history, reflecting the structure of market risks as they have evolved in the past.

The history of companies faced with great technical change is one of a slow death, punctuated with the occasional spectacular collapse. There is little reason to expect the vertically integrated electricity companies to reinvent themselves in the next few decades. Many are doomed to be the dinosaurs of the industry's history. The first casualties are already obvious – E.ON and RWE – but they are unlikely to be alone.

Reinvention has its risks too. The DRAX example stands out. Further back there is a spectacular example of a reinvention that went wrong – Enron. Here a conventional utility turned itself into a trader, excited by the new markets and financial instruments that grew on the back of liberalization of both the financial sector and energy. Enron used criminal methods to hide its risk exposure and it went bust.[13] It was another example of not understanding the risks it faced in entering territory it did not have a deep understanding of, and in a context in which the new commodity traders had begun to establish a distinct and separate trading expertise. Just because market designs change, this does not mean that existing players will be good at the new ones.

CHAPTER 11

The new energy markets and the economics of the Internet

There never has been much of a 'competitive market' in energy. Its history is one of planning, monopoly and vertical integration. The myth of great capitalistic enterprises should not be confused with open competitive markets. The twentieth century saw the big oil and energy utility companies replace markets through the full vertical chains, selling to customers and buying upstream. In the oil case, the only bit of market was when the companies sold their oil to customers, and even then they monopolized the filling stations, leaving little real choice. Industry typically bought energy on longer-term contracts. Big electric utilities typically captured the whole supply chain, leaving nothing or almost nothing to markets. Coal was mined and sold on contracts to generators, which then supplied to monopolized customers. In Europe, this often happened wholly through state-owned enterprises. In China and many resource-rich countries this still happens.

Companies with long-term contracts took the place of competitive markets because markets failed to provide deep, liquid and transparent trading, which could allow risk to be hedged. They also failed because the IT needed to process a high volume of transactions was noticeable by its absence. Finally, once the monopolies were in place they kept markets and competition out. Energy could be driven by technocrats who knew better than markets.[1]

Market trading developed from the 1980s for several distinct reasons, and by 2015 there were wholesale electricity, gas and oil markets on national, regional and global scales. The commodity spot markets we are familiar with now work because there are marginal costs to reflect the value of a barrel of oil, a kilowatt-hour of electricity, a tonne of coal or a cubic metre of gas.

Having grown from nothing to a vast global trading activity, with major new companies like Vitol and Glencore building energy portfolios, the great trading bonanza is beginning a very gradual tailing-off because the energy future is becoming electric, and electricity is increasingly zero marginal cost. Now it is about fixed-priced contracts and fixed-priced auctions. The name of the game has changed, and companies now need to reshape themselves to reflect the risk transfers this new contractual world implies and the evolving nature of energy markets in response to the major changes in technologies set out in Chapter 3, including the emergence of smart data and the new carbon trading.

The future company winners will need to redefine the economic borders of their companies to reflect these developments. The questions range from whether an oil company and an oil service company are different in kind; whether it makes sense to own assets; whether vertical integration works any longer; and the extent to which financial markets can take on the core equity risk components. As we have seen in the previous two chapters, this will be an existential challenge for the big energy incumbents, and one they are unlikely to survive. In these new, very different markets, there are unlikely to be independent oil and gas companies and electric utilities as we know them. First, however, the big energy incumbents need to understand the new markets.

The coming of commodity markets

Prior to the OPEC shocks in the 1970s, the oil companies essentially managed the oil market, internalized it, and made sure that the price was stable. The main activities were organized through contracts, passed down the vertical chain and sold on to customers. To the extent that there was a 'market' it was very much an insider's job. This could be seen by the lack of entry at the various points of the vertical chain. Even at the retail level, would-be competitors had in effect to buy from the incumbents, who were also their rivals.

This cosy oligopolists' world worked reasonably well for both the companies and the major economies, but it did create a serious problem. With long-term contracts there needed to be some sort of reference price against which contracts could be written. By the 1970s this problem was becoming increasingly apparent, and the two main oil indices we know today were designed to address this difficulty: Brent Crude and West Texas Intermediate (WTI). There are of course a large number of different types and locations of oil (and these two indices not only differentiated between Brent in the North Sea and WTI in the US, but also reflected the different types), but what mattered is that most could be referenced back with premia and discounts to these indices.

What transformed these indices into the basis for modern global commodity markets was IT. Few traders have any recollection of what it was like before there were computer terminals and screens to work from. Trading in the pre-IT era was a world of physical presence, where buyers and sellers met in the same place to do their deals. They were select events, and who could trade what with whom was limited. The traders could be genuine insiders, keeping the rest (and especially entrants) out. The scope for rigging these markets to suit the incumbents was large, and it is no accident that entry from outside the industry via these markets was very limited. Auctioneers, specialist stockbroking firms and trading institutions governed the way the markets worked, and prices were posted – scribbled on boards – and the participants often behaved as if in a bear pit.

IT changed all this. Now information was widely and freely available. It did not have to be collected individually and it was much harder to keep it private. There was now a digital trail. Common, abundant information, at the touch of a button, undermined the special and private knowledge of the physical trader. More indices could now be constructed, and with a dose of deregulation. In theory anyone could trade from anywhere with anyone. The 1980s and 1990s saw the rise of the trader, making fast money and living the lifestyle to go with it. In practice, this did not stop further scandals, as the incentives to use private information and to collude remained as strong as ever. But for energy markets, it made trading a much more practical possibility and, importantly, it allowed for a closer tangency with finance.

Nowhere was the new IT more apparent in energy markets than in electricity. The modern electricity 'pools' and wholesale markets are the product of IT. Instead of a planner scheduling in power stations to create a cost-based merit order, using regular telephone calls to station managers to ascertain the state of particular power stations and their readiness to generate, now power stations could bid in prices, and the system operator could translate those prices into the production schedule. Although there still remained the need for a back-up override to the markets (someone still had to be in charge), the role of the system operator had moved from active dictator to a more passive responder to market prices.

As IT developed there seemed no limits to markets, and the roles that markets could take over from the planners. More and more players could be brought into the process of determining prices. These could include speculators who had no direct interest in the outcome. Anyone can buy electricity contracts, and traders do so with the intent of selling them on at profit, with not the slightest concern about the electricity system. The great trading houses referred to above underpinned this explosion of players.

Carbon markets follow this pattern. The EU ETS has been a boon for financial institutions: lots of players with no interest in climate change or the decisions about emissions from particular power stations have got into the game. Indeed, and unsurprisingly, fraud has been a feature of the carbon

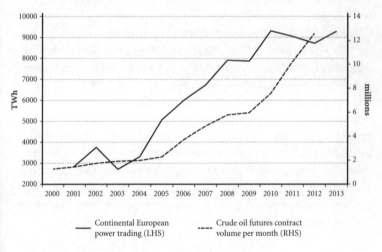

Figure 11.1 Growth of commodity and power trading, 2000–13

Note: European power trading represents measurable power trading volumes at exchanges and brokering platforms (over the counter) in Spain, Scandinavia, Netherlands, France and Germany.

Sources: RWE, 'Power trading: Continental Europe volumes 2001 – 2013', http://www.rwe.com/web/cms/en/403786/rwe/press-news/specials/energy-trading/growth-in-electricity-trading/; Dunn and Holloway, Reserve Bank of Australia, 2012, http://www.rba.gov.au/publications/bulletin/2012/sep/8.html

markets from the outset. As with liberalized commodity markets more generally, the temptation to cheat has been an enduring feature of the EU ETS, and corruption cases have accordingly been all too common.[2] The fact that the commodity – carbon – is practically invisible and diffuse means that monitoring and data have to be built into the market structures, and they have frequently been found wanting.

The big change in energy markets over the last two to three decades that has come with IT has been the entry of the financial markets and the financialization of energy. It has become an asset class in its own right, alongside the other commodities that have followed a similar financialization path.

There have been many consequences, not all of them good. But one of them lies in the ability of financial markets to hedge the sorts of risks that previously could be managed only by vertically integrated oil companies and electric utilities. Instead of internalizing and using long-term contracts to pass down the upstream sunk and fixed costs, now the supplier could buy a futures contract and hedge the price forward. In other words, forward markets could provide a fixed price – the price that a long-term contract would have set. It meant an end to the need for long-term contracts, and hence an end to a core building block of twentieth-century energy companies.

This was an incredible advance, and it meant that liberalization and competition could work, at least in theory. It was no longer essential to own each bit of the vertical chain. Entrants could come in at the supply end, without needing power stations. This was also the case for generators: futures markets could provide them with a fixed price for their output, and hence they did not need end-customers. The same also applied to the purchase of coal and gas: the new spot markets would provide the benchmarks for futures trading and this would hedge their risks. It was but a small step to energy derivatives and a pyramid of increasingly complex financial instruments, ostensibly to better manage risk, but in practice as a speculative opportunity.

The risk problem: sunk costs, switching and spot prices

That, at least, was the theory and it was also the expectation of the more radical liberalizers and promoters of atomistic competitive markets. It would spell the end of vertical integration, and the great planning organizations that energy companies had become.

The regulators proactively encouraged this process. OFGAS, the first energy regulator in Britain, started forcibly undermining the long-term contracts that British Gas had signed to underpin the development of North Sea gas, and pass through the sunk and fixed costs. Once liberalization meant that customers could switch, any of these contracts that had become out-of-the-market would be stranded. A further example was provided by nuclear power, which relied on customers taking on the costs over long periods. If large users of gas and electricity could switch from companies with these higher costs to cheaper spot alternatives, there would be little point in investing in further nuclear power stations or gas developments. Just the possibility that this might happen in future would be enough to deter investors. There would in this world be no Hinkley Point.

In the short term, it did not much matter: in the late 1990s there was a glut of gas and a surplus of electricity-generating capacity in Britain, and later in much of Europe too. In periods of excess supply, competition and switching force prices down, without any consequences from the loss of investment because there is no need for investment. Yet the problems did not go away. They were just masked by the peculiar economic circumstances of the last few decades. The new financial markets undermined the existing players, undoing the implicit or explicit contracts between customers and producers of the old vertically integrated and monopoly worlds. The result was that the foundations for future investment could no longer be relied upon, both generally and for all the new low-carbon technologies. In time, the state would have to step back and reintroduce long-term contracts through direct contracting.

Financialization did not solve the investment problems in oil markets either.[3] The futures markets are still very thin as soon as the time horizon stretches beyond about one year. Oil markets have never relied on hedging to underpin long-term sunk and fixed costs, and the NOCs, with over 90% of the oil reserves under their control, have never let it happen.

The companies have taken the risk, and investors in turn take the equity risk. If you buy shares in Exxon, Chevron, Shell or BP, you will in part be taking a punt on the future price of oil. That can be seen in Figure 9.1 showing the share price reaction to the falls in oil prices since late 2014.

Prior to the 1970s, these companies would have relied on a stable future price of oil. Why? Because they fixed it. Then, the price depended on OPEC, and in particular on Saudi Arabia, and whether collusion worked or not.

Although the oil companies could still balance their businesses from extraction to refining and then retail, the oil price risk became equity risk in the companies and the share prices followed this path – as they have since the oil price collapsed so dramatically at the end of 2014.

As described in Chapter 1, what has contributed to the collapse of the OPEC model has been shale, and a little-noticed feature of shale is its relation to the financial markets. Unlike conventional oil, shale uses rigs which can bring on production very quickly, and in particular within the timeframe of the futures markets. Shale does not need the long-term contracts that Big Oil does because it can turn investment on and off as it sees fit. It is the new short-term swing producer. So shale oil can hedge for the short run; beyond the short-run hedging period, it can take its rigs off the system if prices stay low, and bring them back later on.

The problem that conventional oil faces is one that producer governments face too: they cannot hedge their oil production into the relevant future timeframes. But they have a couple of options. First, they can engage in bilateral long-term contracts. Think of Russia–China and gas, or Sudan–China and oil. These are deals outside the market, and they are not subject to financial intermediation. Second, they can run wealth funds and increase or deplete these funds inversely to the oil price. Thus Saudi Arabia (and many other producers) ran down its sovereign wealth fund in 2015 to offset the losses from the fall in oil prices. This works as long as the assumption is that the price falls in 2015 are temporary, but not if, as argued in Chapter 1, they might be permanent. As Saudi Arabia has realized, the only remaining option is to work out how to live without the 'addiction to oil', and to do so having so far squandered most of the proceeds for the benefit of current generations.

The implication is that whilst the development of financial markets, enabled by IT, has changed energy markets and brought short-term hedging to the fore, it has not fundamentally altered the long-term risk-assignment problem. Indeed, in one sense financial trading in energy has made matters worse. Trading can exacerbate price movements, creating bubbles which, when they burst, disrupt the industry. It may well be that when the history is written, the great financialization of energy markets, to the extent that it happened, will be seen as the product of the specific historical context, and the world will now shift back to a more direct, and political, way of handling the risks.

As already noted, speculative bubbles have been endemic in financial markets since even before the tulip mania hit Holland at the end of the seventeenth century and the South Sea Bubble burst in the early eighteenth century. Bubbles arise when assets are owned not for their primary physical purpose but as finance instruments traded to make speculative profits. Until modern IT and modern financial markets, speculation was largely on the upside: it was hard to bet on prices falling, and hence short sell.

The commodity market speculative game is relatively recent and less studied than the stock market version. In the latter, a huge literature has grown up around the crash in 2000, and the subsequent real estate speculative bubbles as people bought into housing and provided mortgages not on the basis of the value-in-use of the house, but simply to make capital gains.[4] The oil price peaking at $140 per barrel in 2008 has been argued to be one such bubble. A price fall of over 50% in 2014–15 is sometimes explained as the consequence of such a bubble bursting.

The problem has always been to explain why trading at prices that are 'speculative' and above the underlying values could take place. Why would markets allow speculative bubbles to develop, and why do asset holders not sell when prices exceed the equilibrium value? In other words, the question is: why do speculative bubbles ever arise, and why do they persist?

There are several possible answers. The first is the easy one: traders do not know where the equilibrium lies and hence will make what turn out to be *ex post* mistakes. They might, for example, think that China is going to go on growing at 7% GDP per annum, and it may turn out otherwise, as discussed in Chapter 7. The problem with this explanation is that it can explain why prices deviate from the mean, but not why they might deviate *systematically* on the upside: why, therefore, there may be an optimism bias or, as Alan Greenspan, former Chairman of the US Federal Reserve, described it, 'irrational exuberance'.

A more convincing argument derives from the 'beauty contest' example that the economist John Maynard Keynes provided.[5] A speculator might be less interested in the fundamentals than in what the other speculators think. In the beauty contest case, the judge might focus on selecting what was expected to be the choice of the others. The task is to work out what the average person expects to be the case.

This gets us away from a simple linkage of the price to the fundamentals. But there is an obvious flaw: a speculative bubble is rational provided that no one else thinks it is a speculative bubble, or for as long as everyone thinks it will last before bursting. The perfect trick is to still buy above the equilibrium price, but to jump ship before the others, and just before the bubble bursts.

Unsurprisingly, the evidence is far from clear on such bubbles in energy markets, or indeed generally. More information might be expected to even them out. For example, satellite data on coal stocks and ship movements might give a better indication as to what is going on in China. Information means that the facts will be out sooner, and give a better indication of the true fundamentals and hence the equilibrium price. But information works the other way too: new information gets quickly into prices and hence might make them volatile. All existing information is integrated into markets and market prices, and so it is only new information that moves prices.

The impacts on markets can be seen: as financialization and trading have grown, so has price volatility, and the scale of price movements is much bigger in oil and other commodity markets. Since the financial crisis in 2007–08, the oil price has moved from $100 to less than $40, to $140, to less than $50. It is hard to imagine that fundamentals have changed that much in less than a decade.

Zero marginal costs and zero prices

Just as the financial markets and IT have changed the nature of risk allocation and risk spreading, and liberalization has undermined long-term take-or-pay contracts, the very foundations of the energy markets have been changed by the physical characteristics of investments in electricity generation, as well as storage and the coming of electric cars, as discussed in Chapter 3.

Zero marginal costs undermine spot wholesale markets. There is no value in energy to trade. The action moves to capacity markets, and in a fundamental sense back to the past. Zero marginal cost energy means that customers buy *access* to supplies, not the supplies themselves, which are in principle free. As we have seen, this is the economics of the Internet.

A lot can be learned from the economics of the Internet for the future of energy markets, and hence trading, and further for company structures and strategies. There is no deep liquid market in voice calls, or markets for individual Internet searches. From Skype to WhatsApp, the notion of zero marginal cost is widely apparent in almost all broadband and communications services, or at least up to the congestion point, signalling new investment. As a result, spot and futures markets are largely absent. Customer charges may be loosely related to volume as a cost-recovery mechanism, but this price is not a marginal cost. Investment is channelled through entrepreneurs, start-ups and stock market flotations: in other words, it is all about equity and has little to do with hedging and speculating on future prices, whatever they may mean in this context. Companies raise significant funds as equity investments, and the speculative game is all about these initial investments and subsequent stock prices.

In this world, resources are still scarce and there is a resource-allocation problem. There can be auctions for scarce resources, like spectrum, and in principle these can be traded. But it is not the commodity itself that is being traded. To repeat: there are no marginal costs. This zero marginal cost model carries over to energy, or at least to the electricity part, and it is electricity that will gradually drive everything else. The market here is for contracts and capacity, not units of energy.

Capacity and FiT auctions: the case for simplification

It order to develop markets for fixed-price contracts, there needs to be an initial definition of them. Whilst in theory anyone can strike fixed-priced contracts, and indeed these traded contracts effectively hedge the wholesale market through the use of financial instruments like contracts-for-difference, the setting of FiTs and capacity contracts differs in two respects: guaranteeing a wholesale price, and guaranteeing a capacity price for availability.

First, FiTs fix the strike price and hence guarantee the price for the particular output. It is guaranteed if and when the plant generates. It is almost always above the expected wholesale price, though this was not the original intention. As noted above, the mainstream political and company view was that oil and gas prices would rise sharply, and that these would push up electricity wholesale prices substantially. As a result, the FiT prices would, after an initial period, be below the market price. The renewables would prove cheaper than fossil fuel-based alternatives.

This begs the obvious question that if the wholesale price is expected to rise, why could markets not yield these sorts of fixed-price contracts? Why does the government need to provide and guarantee them? The FiTs might be a response to the market failure of futures not being determined far out enough in sufficiently liquid or transparent markets, or they might be a subsidy. In practice, few now expect wholesale prices to rise, and because of zero marginal cost technology, many expect them to fall further. Thus the FiTs are argued to be a subsidy.

Yet this is not entirely right. For if there is going to be more and more zero marginal cost generation, and therefore wholesale prices are going to fall, then the full cost of electricity is not going to be recovered in the wholesale market. If price is related to costs, then this price has to recover the fixed and sunk elements too. The SMP we encountered in the previous chapter may recover the system SRMCs, but it is not going to recover the fixed and sunk costs. These are additional.[6]

The economic problem of the zero marginal cost plant is that its costs are fixed and not variable. What it requires is revenue to recover the fixed and sunk costs – the physical capacity these investments bring to the system. These plants therefore require a capacity contract – a payment for being on the system and being available to generate. The objection is that these plants do not then have an incentive to actually generate. But this is not valid: it is for the system to call these plants to generate, and if they cannot deliver then they should not receive their capacity contracts.

This problem of insufficient revenue from the wholesale market is already widely recognized in respect of the plants with positive marginal costs: the coal and gas power stations. Here the falling wholesale prices result in an SMP that is too low to trigger new investment since the zero marginal cost plant has cut away the wholesale price and, at the same time, rendered these plants intermittent too. They cannot rely on running at baseload. So they get a lower wholesale price and for shorter periods, and the total of the revenue (between the SMP and their own SRMC) cannot cover the full fixed and sunk costs. By applying capacity contracts only to new plant, or old plant which would otherwise close (in other words, to the entry and exit points), the bulk of the existing fossil fuel power stations get expropriated by the zero margin cost plant coming onto the systems and depressing the wholesale prices. This, as we saw in the previous chapter, is precisely what has happened to RWE and E.ON.

Given that the capacity contracts are payments for being available on the system – being willing and able to generate – there is no theoretical objection to bundling the FiTs and the capacity contracts together into a single capacity auction. A central buyer – and there has to be one to take the system view – peers into the future and estimates the required future capacity. It then auctions some of this requirement on a rolling basis, a bit like a central bank issuing gilts on a rolling basis to cover the future borrowing requirements of the government. It does not have to offer every-thing in one go, but rather make adjustments as it goes along.

This new energy market would be a *single unified capacity market*, grad-ually taking up the strain from the wholesale market, which would as a result wither away as a source of revenue to generators. The central buyer would auction contracts on a rolling basis, and a secondary market in these contracts would develop. The key point here is that the simpler the contract design, the greater the liquidity and transparency of the resulting market, and hence the lower the cost of capital, as the risk is spread across financial markets.

Simplicity is hard to achieve because of the political temptation to pick winners. Politicians like to favour particular technologies, to take views on future prices (and, as we have seen, they have done this with spectacular incompetence over fossil fuel prices), and they like to micro-manage the security aspects. Patronage is at the heart of any political system, and paying tribute to vested interests that capture voters, ministers and regulators is correspondingly at the heart of these interventionist models.

If simplicity is a virtue in market design, not least to minimize the sorts of capture described above, then the various dimensions of the market need to be brought within a common framework. When it comes to carbon, the best and most efficient way of meeting the targets is to set a carbon tax, and allow the tax to vary according to how well or badly the targets are being met, a bit like interest rates were supposed to be varied to meet inflation targets. When different technologies bid into the single capacity auction, the carbon target would be endogenous to the bids because the investors would internalize the expected carbon price. Less efficient, but workable (and more likely), would be for a carbon permits market to determine the carbon price, and then this could be similarly endogenized. But if politi-cians continue to pick winners, and treat each technology differently, the result will inevitably be a high cost for meeting a given carbon target, and also a fragmented contract auction with the resulting fragmented secondary contract markets. (The argument that these winners are infant industries which need to be helped to gain competitiveness is a separate one for an R&D policy and a set of innovation subsidies.)

If the carbon element is taken care of through the carbon price, zero marginal cost generators pose a further problem in creating a single auction and market. They are intermittent. But since intermittency imposes additional costs on the system, the obvious efficient solution is to require all plants to bid *firm* power, and hence commit to delivering electricity at the required time. The result would be that intermittent plants subcontract with providers of peaking capacity to address the intermittency, and would have a very strong incentive to encourage the development of batteries and storage, flexible generation and demand-side responses to help this along. Then all contracts would be on the same basis.

This requirement also neatly solves one element of the timing problem for capacity contracts. Some technologies deliver small amounts of power very quickly at peaks. Such peakers are required when there is intermittency. They would find their niche through the secondary market for the intermittent bidders. Other projects take a long time to come on the system. Take a new nuclear plant. If it wins a contract now, it may not generate for a decade. And then, unlike most other technologies, it may generate for up to sixty years. It is argued that such projects are 'special' and need their own protected guarantees on price (and waste and decommissioning).

Again the single auction deals with this problem. These projects could bid for the period they are going to generate for. They will bid for contracts starting in, say, ten years' time. The very fact that they are included in the market helps to create a longer-term forward market in capacity and the liquidity that this brings. The objection is that the generation starting date is uncertain. But as with the intermittent renewables above, this is a cost they impose on the system and they should be required to sub-contract to those that can provide the capacity if they fail to complete the plant on time.

A unified single capacity market is the way in which an increasingly decarbonizing energy world could be contextualized. It may not of course happen: governments may continue to plan the market internally, through direct contracting with each technology, and even taking the nationalized industry route that dominated the twentieth century. Planning through direct contracting and state ownership of investment are examples of the same sort of model: both determine who invests in what, and what they get paid. State control does not require state ownership.

Yet the failures of state control are obvious and will become increasingly burdensome. The Germans have saddled their consumers and their economy with some of the highest electricity prices in the world by picking first-generation solar panels for their cloudy skies, and lots of wind farms. Britain is discovering the scale of its subsidy commitments. Across Europe – the global leader in picking low-carbon 'winners' – the subsidies are being rolled back and governments are increasingly resorting to auctions. It will take time for the new markets to mature, but it is likely that they will – because they are cheaper, and because customers cannot easily afford to pay for the costs of the losers that governments have so often picked. Most importantly, the capacity market route goes with the grain of the fixed- and sunk-cost structures. These do not require politicians to predict future fossil fuel prices (and make mistakes on a grand scale like the belief in peak oil and gas).

Knock-on impacts on gas markets

The development of capacity markets in electricity, and the move from wholesale to capacity prices, has profound implications for the gas industry. Gas (and oil) is based on energy and not capacity costs and prices. Long-term contracts are about prices: they fix prices to ensure that fixed and sunk costs are recovered.

As the market for gas has grown to generate electricity, the marginal demand for gas has become partly the marginal demand for gas-fired power stations (with heating largely given). It is a relatively new market: as noted, it was illegal to burn gas in power stations up until 1990 in Europe and the US. Gas was regarded as a premium fuel to be reserved as a petrochemicals feed-stock. The early dash-for-gas in electricity was to fuel large CCGTs, which would then run baseload. However, the decarbonization strategies have promoted zero marginal cost generation, and this in turn has rendered the gas CCGTs intermittent too (and in some cases forced them to be moth-balled). Small open-cycle gas turbines have proved more competitive in this new flexible-generation world.

The problem for gas companies is: how to contract for intermittent customers, as the CCGTs have become? The answers are several. First, the gas contracts now need to have a capacity element: to be available if and only when demanded. So the gas companies have a risk associated with the capacity market, and in effect become part of that market. Second, like the intermittent generators, they have a problem of backing up their own inter-mittency – to make sure they have the gas when demanded, but also to have a market for surplus gas when not demanded.

The requirement for firm bids in the electricity capacity markets encour-ages the development of batteries and other ways of storing intermittent electricity generation. On the gas side, the problem will be reflected in the bundling of contracts with CCGTs: the bigger the portfolio, the smaller the impact of particular intermittent demand. In both cases, the economics of linkages between electricity and gas systems will be improved too.

New markets in traded regulatory asset bases: back to the utility model

The coming of capacity markets will have much in common with the digital communications markets, where marginal costs are also zero and the pricing is about capacity constraints. Thinking about energy markets in this way suggests wider market couplings. One such link is with utilities and their underlying regulatory asset bases, rather than through the commodity trading discussed above.

A utility has at its core a problem that is shared with zero marginal cost generation. It has a big gap between its marginal and average costs, giving rise to a time-inconsistency problem. If it invests in, say, a new gas pipeline or a transmission link, it faces an initial capital cost. But once built, the costs of using it are close to zero (zero marginal costs). Hence *ex ante*, it needs to be guaranteed a price to recover its average costs, but *ex post* it will operate as long as it meets its SRMCs. It can be expropriated for the differ-ence. Rate of return regulation, public ownership and a guaranteed regula-tory asset base are some of the solutions that have been applied.

The utility regulatory regime differentiates between three types of project risk: operating the existing system, carrying out investment projects, and the time-inconsistency risk. These categories apply equally to power stations. A wind farm costs a bit to run, it has major project-development risks, and it is exposed to *ex post* marginal cost pricing failing to recover its average costs.

In terms of risk allocation, the first is a management task that requires an operating margin, the second is an equity risk, and the third is a debt risk, since management can do nothing about it. In the energy commodity model, the first is about the wholesale price (covering the marginal costs), the second is about the equity markets, and the third is about securitizing the projects once complete.

It is this third element that is most akin to the regulatory asset bases in the utilities. If an infrastructure fund buys a completed wind farm, and if the operations are contracted out (with guarantees to take over the risk of equipment failure), then what is left is a regulated asset base, but in this case one covered by the capacity contract or FiT rather than a regulator offering a direct asset value guarantee. In an important sense the electricity industry has converged with the utility sector, as has already been recognized by some early and leading infrastructure funds. The cost of capital, essentially a cost of debt, has as a result started to converge with that for the utilities.

To this is added the utility elements of the energy sector itself, and these are primarily oil and gas pipelines, LNG terminals, electricity transmission systems and large-scale oil, gas and electricity storage systems. What are left to the mercy of the energy wholesale markets are the oil, gas and coal commodities, and this, as we have seen, is likely to be a declining future.

Markets of the future

The energy markets of the future are being driven by the three big changes. The commodity super-cycle is over and lower oil and gas prices will impact on investments from oil E&P right through to renewable electricity generators. Decarbonization will drive a wedge into the wholesale electricity markets, and these will ricochet upstream into gas markets. Finally, new technologies will reinforce the move away from marginal cost pricing, and render the energy systems much more like the Internet – a system with access charges rather than a commodity market.

It is ironic that these changes come just as the financial markets have completed their innovations in the way commodities are traded and have developed ever-more sophisticated ways of hedging commodity risk. They have developed financial instruments that look less and less important to the future energy markets.

Even this role has been exaggerated. The traded energy markets are largely short-term, and the market itself is increasingly driven by states, not market forces. The oil market is coordinated by state-owned oil companies, which blur the politics and the economics, and are often opaque at best in accounting terms, and sometimes corrupt. In electricity, the state is back: the single buyer has taken over in place of the atomized and liberalized market. Systems need someone to be in charge, and they need system operators.

The new market structures, built around capacity, will take shape slowly. Politicians will find it hard to admit that the picking of winners is actually a loser's game and that losers pick governments. Early capacity markets are riddled with complexities thanks to this micro-meddling. Yet the problem with such inefficient state interventions is that the consequences cannot be avoided. Costs tend to be higher and voters eventually notice, even in Germany. The broader markets in goods and services incorporate energy costs, and inefficient energy policies result in a loss of competitiveness. Voters tend to revolt too.

The path may be hesitant and take time, but the emergence of a new energy market model has an element of inevitability. The fall of commodity prices and decarbonization policies will push in this direction, but it is the arrival of new solar technologies, batteries, storage and electric cars, and smart demand-side technologies that will eventually kill off the conventional wholesale market model which has dominated the energy markets for the last century.

Conclusion

In his authoritative study, *The Rise and Fall of American Growth*, Robert J. Gordon argues that the twentieth century was blessed with the coincidence of the 'Great Inventions' of the late nineteenth century, and that these opened up the unprecedented economic growth from around 1870 through to 1970.[1] At the heart of this great explosion of invention lay energy, and in particular electricity. It brought electric light, electric elevators (which enabled tall buildings), electric refrigeration (which transformed food supplies), and electric machines in the factories and especially in households. Equally momentous were the internal combustion engine and the subsequent emergence of mass transportation.

If it was energy that transformed human prospects after millennia in which nothing much happened in economic growth terms, Gordon thinks the great growth period is now over. Technical change since 1970, he argues, has been slow, and incremental in entertainment, communications and IT. This fits with the pessimism of our age, in the shadow of the economic crash in 2007–08. In the period from around 1970 to the early 1990s, he is right. It did take time for ideas to translate into products and services, and it has taken a while for IT to evolve from a useful add-on to a general pervasive technology. But in the past two decades it has taken off, and the new information technologies are proving every bit as radical as those in electricity a century before.

Over the coming decades, the impact of the next wave of Great Inventions will be felt, and they share with those of the period after 1870 a core focus on energy. Although electricity and the internal combustion engine transformed our world in the twentieth century, they did not solve all our energy problems. The legacy of the great conventional energy century is very much with us in the carbon stock in the atmosphere and the terrible political consequences of the resource curse that oil has brought to most of the key producers (but not the US).

The key features of our energy transformation now are speed and the multitude of *simultaneous* advances. Though the oil companies cling to their belief in the staying power of the fossil fuels, and climate policymakers think we should invest heavily in the current conventional renewables on the grounds that the new technologies cannot be expected until the second half of the century, both are likely to be making big mistakes.

The speed with which shale oil and gas, and the fracking that made it possible, transformed the oil and gas global markets is extraordinary. In the space of less than ten years, US shale oil added 3 mbd to its output, and US shale gas became so abundant as to switch the expectation of large-scale imports to the first exports, and to encourage the reshoring of petrochemicals back from China to the US. Within the next decade, the US can contemplate a rough North American energy independence. Even with the massive demand growth caused by the Chinese transition, shale has punctured the great commodity super-cycle. In the process, it has – again within a decade – ended the US's economic dependency on the Middle East, and Saudi Arabia in particular, and given a considerable boost to US manufacturing against the relentless pressure that China has brought to bear with its cheap labour.

Technical change has been revolutionary in fossil fuels – yet it has been a revolutionary change within the existing set of fossil fuels. It is harder to make this argument in IT. Here, the transformation has so far been incremental. Mainframe computers morphed into desktops and laptops, and fixed-line telephones morphed into mobiles. But these are not the only or even key drivers. It is the Internet and the digitalization of almost everything that is only beginning, but already accelerating at a rate which looks to have no historical parallels, though it does concentrate on similar areas to those that so excited Gordon about the period after 1870: energy, transport and electricity. Everything that is digitalized is electric, including this time technologies relating to transport.

The technical change in energy is not just about one dimension, as, for example, was the development of the coal power station in the late nineteenth century, and then the gas and nuclear power stations in the mid-twentieth century. This time it is about generation *and* storage *and* distribution *and* supply – the whole vertical chain – *and* the merging of transport with electricity. In the background lies the transformation of manufacturing *and* services *and* households. The new information technologies are transforming manufacturing and services with robots, 3D printing and AI. Crucially, these are not discrete inventions: they are intimately connected, dependent on each other and the mass of data now collected, analysed and deployed. Machine learning is a first stage, but super-intelligence goes beyond the substitution for human labour and into a whole new world of activities that involve computer learning. Together these great changes, all coming together in the last two decades, are every bit as momentous as those Gordon identified in America after the end of its Civil War.

What will this new energy world look like? We cannot know in any detail what these changes will bring. The correct answer is that it will be surprising. There are a host of possible energy futures. Some think we will have cracked nuclear fusion, opening up unlimited supplies. Others see a world of continuing fossil fuel dominance of the energy mix. Although we cannot be sure, both of these energy futures are unlikely. The former just takes a long time. The latter requires us to suspend our faith in technical progress, and indeed put some of the new technologies back in their boxes.

But we are not entirely ignorant. More likely is that all three of our predictable surprises will have played out. Fossil fuels will be on their way out, and many of today's big energy companies will be unrecognizable – either because they have gone bust, or because they have invented a post-fossil-fuel world for themselves. The big vertically integrated electricity utilities will have gone.

The Middle East and Russia will be very different too. Both will have had to come to terms with the decline and then death of their great golden geese. Saudi Arabia will no longer be a spectacularly rich nation, and it may well have come to blows with its neighbours in Iran and possibly Iraq by then too. Indeed this might come quite soon. The endless cycle of post-Ottoman violence will still be playing out, but now on a declining resource base rather than the burgeoning one that has characterized the Middle East since the end of the First World War. Things will no longer be getting better year after year: they will be getting worse.

Russia will have reverted to its 1990s economic status as a middling country, though not without more chaos around its borders. War, overt or more hybrid, may have escalated beyond eastern Ukraine and the Baltics may have experienced the full anger of the wounded bear that Russia will have become. It will be further pressurized by Chinese immigration in the east as the Han people push on still further from their heartland in the North China Plain and their settlement of Tibet and Xinjiang provinces into eastern Siberia.

These companies and countries will be the casualties. The resource curse in reverse applies viciously to both companies and countries. The future of the great oil and gas producers looks bleak: decline is rarely an easy political path, and it will be especially tough for those countries with young populations and diminishing largesse to buy off revolts. But before we get carried away with the gloom, there is time to do something about it. The winners in the great energy transformation, especially the Americans and the Europeans, have the opportunity to help these transitions. If not, they can at least avoid making them worse.

Even if the Middle East and Russia descend into more brutality and violence, neither the US nor Europe needs them. The energy transformation means an end to dependency. The Europeans will no longer be exposed to the threats and bullying by Russia to cut off its gas supplies. The Americans will no longer need Saudi Arabia. Even the Chinese may find that their geopolitical strategy to control the South China Sea and the Strait of Malacca, and to project power into the Indian Ocean, is no longer necessary to secure their energy needs. The new information technologies may also bring an end to the great growth of world trade, as more localized production closer to the final consumers is enabled.

Most likely, China will have other things to worry about as its transformation ends, its population ages, and the bargain of absolute Communist Party control in exchange for economic advancement breaks down. In our energy future, what exactly is the point of all those oil tankers and LNG ships sailing through the Strait of Malacca and on through the South China Sea?

The transformation of our energy world will have brought great benefits to offset some of the negative impacts. By 2050 we may be well on our way to cracking climate change, without the assistance of grand global plans like those hatched at Paris. Climate change is a *solvable* problem, but only with the steady march of new technologies, and these have to be electric. Top-down grandstanding has not got us very far. Grand speeches about 'saving the world' from political leaders make good theatre but not good energy policies. The question is whether the technological transformation will come fast enough before serious damage is done by higher temperatures.

Given how slow technical progress has been across the energy technologies, and especially in electricity, in the last century, it is not surprising that the conventional view is that the predictable surprises outlined in this book are all post-2050. This assumption is also convenient for many vested interests. It comforts the oil companies, which see a growing market through to 2050, and hence plenty of miles (and dividends) left on the fossil fuel path. It suits the conventional renewables and supports their arguments for more subsidies now: why not spend lots on current solar and current wind farms, since we cannot afford to wait until 2050 for alternatives? It suits conventional nuclear too: why not subsidize big PWRs now, safe in the knowledge that they have thirty-five years to pay back the investment before the threat of new technologies overtaking them becomes a possibility?

But such technological pessimism is overdone, and dangerous in that it leads to inefficient policies benefiting the incumbents at the expense of entrants. These new technologies are an exciting prospect for entrepreneurs, innovators and scientists, and they are resulting in the emergence of many new companies, supported by tech investors. Most of these new companies will fail, and many investors will lose a lot of money. But not all will. Some will bestride the energy futures. We have already had Microsoft, Apple and Google come from nothing to dominating world markets in their products inside two to three decades. Yet these are just the start. Who will follow? We can't know the names – many of them have yet to be set up. But we do know where the R&D is concentrated, and so again we are not entirely in the dark. When it comes to electric cars and the associated digitalization of transport, the big money is being invested by a small number of players – in existing car companies such as Toyota, Nissan and BMW, and in new players like Google and Tesla. In electricity, there are numerous solar companies, some of which have already gone bust, and many more start-ups and university spin-offs. Total, BP and Shell have all either tried and failed or are attempting to join in. In storage and batteries, there is again the mix of the incumbents and the new. In smart data, meters and decentralized generation and networks, most are new, though some, like Centrica, are trying their hands.

What will we experience of energy in this world? How will our lives be transformed? Although there is much anxiety about the replacement of labour by robots, and about the coming of super-intelligence that we cannot control, the future can only be postponed. It cannot be stopped. These threats to social arrangements and culture will need to be managed, and there will be losers, notably among the unskilled and uneducated, and the great middle classes employed in banks, law and accountancy. Indeed there already are, and with this comes the possibility of an even greater polarization of wealth and income.

But such doom needs to be tempered with recognition of the benefits that the new technologies will bring. First and foremost they offer an end to climate change, and probably the only possible end. Cutting demand for energy is never going to work, and existing renewables cannot size up to the scale of the problem. If climate change is the existential threat many scientists tell us it is, then the great new wave of inventions is a get-out-of-jail-free card – just as the coming of electricity was to the drudgery of what was still a predominantly rural society at the end of the nineteenth century in the US and much of Europe.

Climate change will get solved. But in the process much else will be better. The drudgery of everyday life is different now to a century ago, but much of it is still repetitive, boring and time-consuming. Almost everything about improving human life remains focused on energy – energy for all the new appliances, from household robots to custom-made products close to the home, and the personalized transport that driverless cars offer, especially to the old and disabled who may have a great boost to their mobility. All need electricity. The mantra about saving energy, about it being the best option for all three of the policy objectives – security, carbon and affordability – has never been convincing and could become irrelevant. Not only can energy not be 'saved' (we cannot literally consume it), but there may even be no need with the possibility of unlimited solar (or even nuclear) power.

Think what this would mean. Electricity that is cheap, so cheap that the marginal cost will be zero. It will be all about capacity, not energy. There will be no wholesale electricity price, because there will be no wholesale electricity market. This would be one of the greatest advances for households and has the potential to massively reduce fuel poverty.

These changes cannot be stopped, but this does not mean that governments and policymakers have no role to play. They can slow down the process, and they can obstruct the deployment of new technologies. They can pander to the lobbyists for the current technologies, both fossil fuels and current generation renewables and nuclear, protecting their markets and giving long-term subsidies paid out of household budgets. These vested interests spend large sums trying to capture governments, and they are very successful at it. The politicians and the regulators play musical chairs between their public roles and the private jobs they take.

Creating an energy policy that meets the public rather than these private interests is not easy, but neither is it impossible. Going with the grain of the new technologies, facilitating their deployment through capacity markets, and recognizing that although R&D inevitably has lots of failures, it is generally a much better use of public monies than propping up the incumbent interests – these form the backbone of a sensible energy policy. Better to encourage entrepreneurs and accept the many mistakes than to try to hold back the tide, and in the process wreck the planet.

Endnotes

Introduction

1. See Pfeiffer, A., Millar, R., Hepburn, C. and Beinhocker, E., 'The "2°C Capital Stock" for Electricity Generation: Cumulative Committed Carbon Emissions and Climate Change', Applied Energy, 2016.
2. Jevons, W. S., *The Coal Question: An Inquiry Concerning the Progress of the Nation, and the Probable Exhaustion of Our Coal Mines*, Dodo Press, 2008.

Part One: Predictable Surprises

1. Responding to a question at a US Department of Defense news briefing on 12 February 2002, Donald Rumsfeld stated that: 'Reports that say that something hasn't happened are always interesting to me, because as we know, there are known knowns; there are things we know we know. We also know there are known unknowns; that is to say we know there are some things we do not know. But there are also unknown unknowns – the ones we don't know we don't know.'

Chapter 1: The end of the commodity super-cycle

1. Blas, J. and Flood, C., 'Analyst Warns of Oil at $200 a Barrel', *Financial Times*, 6 May 2008, http://www.ft.com/cms/s/0/70b4ef0a-1b91-11dd-9e58-0000779fd2ac.html?sitee dition=uk#axzz3igHrWU9r. The article states that Chakib Khelil, then president of OPEC, also warned that oil could reach $200 per barrel.
2. Hamilton argued that there are five ways in which the world of energy may have changed for ever: that world oil demand is now driven by the emerging economies; that growth of production since 2005 has come from lower-quality hydrocarbons; that stagnating world production of crude meant significantly higher prices; that geopolitical disturbances held back growth in oil production; and that geological limitations are another reason that world oil production stagnated. Hamilton, J. D., 'The Changing Face of World Oil Markets', NBER Working Paper 20355, July 2014, http://www.nber.org/papers/w20355
3. For a series of quotes, see Helm, D., *The Carbon Crunch: How We're Getting Climate Change Wrong – And How to Fix It*, Revised Edition, Yale University Press, 2013, Chapter 7.

4. It is disingenuous to claim that they are projections, not forecasts. All forecasts are projections. The underlying issue is whether these are based on formal models or informal assumptions. Extrapolating trends are not based on either. See Aurora Energy Research, 'Predictable Surprises: Lessons from 30 Years of Energy Sector Forecasts', 2013.

5. Aurora Energy Research, with quotes from: Reuters, 'Russian central bank prepares strategy for sharp oil price drop', 1 October 2014, http://uk.reuters.com/article/russia-cenbank-oil-idUKL6N0RW1G520141001; *Telegraph*, 'Oil price crash is "fault" of non-OPEC members, says Saudi oil minister', 21 December 2014, http://www.telegraph.co.uk/finance/newsbysector/energy/oilandgas/11306450/Oil-price-crash-is-fault-of-non-OPEC-members-says-Saudi-oil-minister.html; Reuters, 'OPEC's Badri says hopes for oil price revival by end H2 2015', 21 December 2014, http://uk.reuters.com/article/uk-oil-prices-opec-badri-idUKKBN0JZ0J020141221; *Financial Times*, 'BP chief expects low oil prices until at least 2016', 3 June 2015, https://next.ft.com/content/e7888f30-09f8-11e5-b6bd-00144feabdc0; Bloomberg, 'Shell CEO Sees Oil Rising to $70 to $90 in "Long Run"', 30 July 2015, http://www.bloomberg.com/news/videos/2015-07-30/shell-ceo-sees-oil-rising-to-70-to-90-in-long-run-

6. Dan Yergin's book *The Prize* remains the best popular history of the oil industry. Yergin, D., *The Prize: The Epic Quest for Oil, Money, and Power*, Simon & Schuster, 1990.

7. See Terzian, P., *OPEC: The Inside Story*, translated by Michael Pallis, Zed Books, 1985.

8. Carter, J., 'The President's News Conference', 13 February 1980, accessed at The American Presidency Project, http://www.presidency.ucsb.edu/ws/?pid=32928. On Carter's energy policies see Graetz, M. J., *The End of Energy: The Unmaking of America's Environment, Security and Independence*, MIT Press, 2011, Chapter 7.

9. Helmut Schmidt, speaking at the Tokyo Economic Summit Meeting, Tokyo, 28–29 June 1979, accessed at US Department of State, Office of the Historian, Foreign Relations of the United States, 1969–1976, Volume XXXVII, Energy Crisis, 1974–1980, Document 221, http://history.state.gov/historicaldocuments/frus1969-76v37/d221

10. The French President Valéry Giscard d'Estaing stated that: 'The main economic problem is the energy problem: oil supplies, and the securing of these supplies in the short, medium and long term ... Regarding alternate sources, the main ones are nuclear energy and coal. Other alternatives are not yet available. On coal and nuclear energy, we should express a determination to speed up production. We are all clearly concerned about safety, but this should not be an *a priori* condition to further new energy development, because if it is it will delay energy development.' The President of the European Commission, Roy Jenkins's opinion was that 'The long-term trends in oil prices are going up, and we can't avoid oil price increases.' See Tokyo Economic Summit Meeting, 28–29 June 1979.

11. Exxon's 1979 *World Energy Outlook* states that: 'Exxon said that, after adjusting for inflation, the price of Middle East crude oil is expected to rise about 50 percent by the end of the century from its current level.'

12. See Energy Modeling Forum, 'World Oil', EMF Report 6, Summary Report, EMF, February 1982, pp. 2–3, in Gately, D., 'Lessons from the 1986 Oil Price Collapse', *Brookings Papers on Economic Activity 2*, 1986, http://www.brookings.edu/~/media/Projects/BPEA/1986-2/1986b_bpea_gately_adelman_griffin.PDF

13. Initially in the mid-1970s nuclear benefited in the US, until the 1979 accident at Three Mile Island changed the political and public mood.

14. British Prime Minister Margaret Thatcher's view at the time: 'This sharp oil price increase has happened for the second time within a decade. It is a long and short term problem ... We need more nuclear and must convince countries that nuclear is safe.' Ohira Masayoshi, Prime Minister of Japan, stated that: 'Regarding the point on peaceful nuclear energy, I feel that the most reliable, realistic alternative to oil is nuclear energy. We have adopted a course leading to more nuclear energy, and we expect the most of nuclear energy as an oil alternative. With the U.S., Canada and France we have

developed and are moving forward on technical cooperation arrangements.' See leaders'
statements at the Tokyo Economic Summit Meeting, 28–29 June 1979.

15. Meadows, D. H., Meadows, D. L., Randers, J. and Behrens, III, W. W., *The Limits to Growth: A Report for the Club of Rome's Project on the Predicament of Mankind*, New American Library, 1972. For a broader critique of the Club of Rome and its Malthusian underpinnings, see Helm, D., *Natural Capital: Valuing the Planet*, Yale University Press, 2015, pp. 29–32.

16. Hubbert, M., 'The Energy Resources of the Earth', *Scientific American*, 225, September 1971, pp. 60–70. Hubbert was very much of the Club of Rome school. He had been a leading member of the 'technocracy' movement in the 1930s, which held that the price mechanism should be replaced by technocratic decision-making. It is precisely his opposition to economics generally, and the incentive effect of prices, which led him into the errors of peak oil. See Inman, M., *The Oracle of Oil: The Maverick Geologist's Quest for a Sustainable Future*, W. W. Norton 2016, Chapters 1–3.

17. See, McKane, A., Price, L. and de la Rue du Can, S., 'Policies for Promoting Industrial Energy Efficiency in Developing Countries and Transition Economies', Background Paper for the United Nations Industrial Development Organization (UNIDO) Side Event on Sustainable Industrial Development on 8 May 2007 at the Commission for Sustainable Development (CSD-15), p. 9.

18. See Nickell, S., Redding, S. and Swaffield, J., 'The Uneven Pace of Deindustrialisation in the OECD', *The World Economy*, 2008.

19. More parochially, the case of Scottish independence set out by the Scottish government in its 2013 White Paper assumed an oil price of $113. The Scottish Government, 'Scotland's Future: Your Guide to an Independent Scotland', 26 November 2013.

20. Frederick van der Ploeg states that 'Even the mere risk of a political regime switch towards enforcement of the 2 degree cap on peak warming will induce a race to burn the last ton of carbon.' Ploeg, F. van der, 'Race to Burn the Last Ton of Carbon and the Risk of Stranded Assets', Research Paper 178, Oxford Centre for the Analysis of Resource Rich Economics, Department of Economics, University of Oxford, p. 14.

Chapter 2: Binding carbon restraints

1. See Helm, D., Phillips, J. and Smale, R., 'Too Good to be True? The UK's Climate Change Record', 2008, www.dieterhelm.co.uk; and more recently, Peters, G. P., Minx, J. C., Weber, C. L. and Edenhofer, O., 'Growth in Emission Transfers Via International Trade from 1990 to 2008', *Proceedings of the National Academy of Sciences of the United States of America*, 108(21), 24 May 2011, pp. 8903–8, http://www.ncbi.nlm.nih.gov/pubmed/21518879

2. Between 1990 and 2014, greenhouse gas production in Europe fell by 23%.

3. The expansions came in the Helmstedt, Lausitz and Central German mining regions. New pits were opened in the Rhine area.

4. The only other major achievement was to establish a second Kyoto period through to 2020 for the willing. They needed this to keep the EU ETS afloat.

5. International Energy Agency, 'Medium Term Coal Outlook: Market Analysis and Forecasts to 2020', IEA, 2015.

6. See, for example Nielsen, C. P. and Ho, M. S., 'Air Pollution and Health Damages in China: An Introduction and Review', in Nielsen, C. P. and Ho, M. S. (eds), *Clearing the Air: The Health and Economic Damages of Air Pollution in China*, MIT Press, 2007.

7. Reuters, 'India says Paris climate deal won't affect plans to double coal output', 14 December 2015, http://www.reuters.com/article/us-climatechange-summit-india-coal-idUSKBN0TX15F20151214

8. See UNFCCC website for Intended Nationally Determined Contributions (INDCs), 'India INDC to UNFCCC', October 2015, http://www4.unfccc.int/submissions/INDC/

Submission%20Pages/submissions.aspx. See also Climate Home, 'India Considers Emissions Peak 2015–50', 3 December 2014, http://www.climatechangenews.com/2014/12/03/india-considers-emissions-peak-2035-50/

9. Tol, R. J., 'The road from Paris: International climate policy after the Paris agreement of 2015', VOX CEPR Policy Portal, 17 December 2015, http://voxeu.org/article/road-cop21

10. See MacKay, D. J. C., *Sustainable Energy – Without the Hot Air*, UIT Cambridge, 2008. On solar, MacKay assumes an efficiency of just 10%, whereas solar farms can now deliver 15–20%, and this changes their economics somewhat. Nevertheless, the general point remains valid.

11. For an analysis of the issues raised by a DRAX-type of supply chain, see Stephenson, A. L. and MacKay, D., 'Life Cycle Impacts of Biomass Electricity in 2020: Scenarios for Assessing the Greenhouse Gas Impacts and Energy Input Requirements of Using North American Woody Biomass for Electricity Generation in the UK July 2014', Department of Energy and Climate Change, July 2014, https://www.gov.uk/government/uploads/system/uploads/attachment_data/file/349024/BEAC_Report_290814.pdf

12. None of this has stopped over 50% of the EU renewables target being met by biomass.

13. National Energy Policy Development, 'National Energy Policy: Reliable, Affordable, and Environmentally Sound Energy for America's Future', May 2001.

14. See Helm, *The Carbon Crunch*, pp. 97–9; and Knittel, C., 'Reducing Petroleum Consumption from Transportation', *Journal of Economic Literature*, 26(1), Winter 2012.

15. See BBC, 'China's grim history of industrial accidents', 29 January 2016, http://www.bbc.co.uk/news/world-asia-china-35149263

16. Oil is not good from an emissions perspective, and in any event is not a great way of generating electricity, even if some oil-rich Gulf States do use it. In transport, diesel particulates are of great concern. Gas is much better, though still of course a carbon fuel.

17. See Coady, D., Parry, I., Sears, L. and Shang, B., 'How Large are Global Energy Subsidies?', IMF Working Paper WP/15/105, 2015.

18. See Hepburn, C., 'Regulating by Prices, Quantities or Both: An Update and an Overview', *Oxford Review of Economic Policy*, 22(2), 2006, pp. 226–47.

19. See Helm, *The Carbon Crunch*, pp. 186–90. For a more favourable interpretation, see Delbecke, J. and Vis, P. (eds), *EU Climate Policy Explained*, Routledge, 2015.

20. There have been numerous twists and turns in the development of the proposals to reform the EU ETS. See http://ec.europa.eu/clima/policies/ets/reform/index_en.htm

21. Allen, M., Frame, D., Huntingford, C., Jones, C., Lowe, J., Meinshausen, M. and Meinshausen, N., 'Warming Caused by Cumulative Carbon Emissions Towards the Trillionth Tonne', *Nature*, 458(7242), 2009, pp. 1163–6; and also Millar, R., Allen, M., Rogelj, J. and Friedlingstein, P., 'The Cumulative Carbon Budget and its Implications', *Oxford Review of Economic Policy*, 32(2), Summer 2016, pp. 323–42.

22. Haszeldine, R. S., 'Can CCS and NET enable the continued use of fossil carbon fuels after COP21?', p. 304, in Helm, D. (ed.), *The Future of Fossil Fuels: Oxford Review of Economic Policy*, 32(2), Summer 2016, www.oxrep.oxfordjournals.org

Chapter 3: An electric future

1. The British had a policy of using old mine shafts and deep-water trenches for disposal (Radioactive Substances Advisory Committee, Panel on Disposal of Radioactive Wastes Cmnd 884, 'The Control of Radioactive Wastes', HMSO, 1959), and there can be no doubt that nuclear submarine and other debris litters the Arctic seabed.

2. See Leveque, F., *The Economics and Uncertainties of Nuclear Power*, Cambridge University Press, 2015.

3. See www.iea.org/tcp/fusionpower/ntfr/; and also http://www.world-nuclear.org/information-library/current-and-future-generation/nuclear-fusion-power.aspx

4. For a flavour of the issues in the political debate about the renewable subsidies in Germany, see http://www.euractiv.com/section/trade-society/news/germany-to-cut-energy-rebates-for-industry-renewable-subsidies/. The true scale of the subsidies is highly contested, in particular because of the impacts on the system and the losses to the conventional producers.

5. On the processes of innovation, see Lester, R. K. and Hart, D. M., *Unlocking Energy Innovation: How America Can Build a Low-cost Low-carbon Energy System*, MIT Press, 2012; and Fagerberg, E. J., Mowery, D. and Nelson, R., (eds), *The Oxford Handbook of Innovation*, Oxford University Press, 2005.

6. See Pazos-Outón, L. M., Szumilo, M., Lamboll, R., Richter, J. M., Crespo-Quesada, M., Abdi-Jalebi, M., Beeson, H. J., Vrućinić, M., Alsari, M., Snaith, H. J., Ehrler, B., Friend, R. H., and Deschler, F., 'Photon recycling in lead iodide perovskite solar cells', *Science*, 351, 2016, pp. 1430–3.

7. See http://www.nrel.gov/workingwithus/re-csp.html, and in particular the link to video explaining the technology.

8. For a survey of grid scale storage technologies, see Brandon, N. P. et al., 'UK Research Needs in Grid Scale Energy Storage Technologies', http://energysuperstore.org/wp-content/uploads/2016/04/IMPJ4129_White_Paper_UK-Research-Needs-in-Grid-Scale-Energy-Storage-Technologies_WEB.pdf

9. For more background on battery research see publication list at http://pgbgroup.materials.ox.ac.uk/publications.html and http://news.mit.edu/topic/batteries

10. Moore's Law refers to the observation made in 1965 by Gordon Moore, co-founder of Intel, that the number of components per chip would double every year (Moore, G. E., 'Cramming More Components onto Integrated Circuits', *Electronics Magazine*, 1965, p. 4). Ten years later he revised this rate of complexity to doubling every two years (Moore, G. E., 'Progress in Digital Integrated Electronics', IEEE, *IEDM Tech Digest*, 1975, pp. 11–13).

11. See Shultz, G. P. and Armstrong, R. C., *Game Changers: Energy on the Move*, Hoover Institute Press, Stanford University, 2014, Chapter 3.

12. For a survey of the issues, see Mayer-Schönberger, V. and Cukier, K., *Big Data: A Revolution that Will Transform How We Live, Work and Think*, John Murray, 2013.

13. See Abraham, D. S., *The Elements of Power: Gadgets, Guns and the Struggle for a Sustainable Future in the Rare Metal Age*, Yale University Press, 2016.

14. The Bill and Melinda Gates Foundation funded this application, given the strength, flexibility and thinness of graphene. See 'Bill Gates condom challenge "to be met" by graphene scientists', 20 November 2013, www.bbc.co.uk/news/uk-england-manchester-25016994

15. Brynjolfsson, E. and McAfee, A., *The Second Machine Age: Work, Progress, and Prosperity in a Time of Brilliant Technologies*, W. W. Norton, 2016.

16. See MIT Technology Review, December 2015, https://www.technologyreview.com/s/544421/googles-quantum-dream-machine/

17. On the wider implications, see Ford, M., *The Rise of the Robots: Technology and the Threat of Mass Unemployment*, Oneworld Publications, 2015.

18. See http://www.eia.gov/todayinenergy/detail.cfm?id=25632

Chapter 4: The US

1. The 1823 Monroe Doctrine declared that the US sphere of influence was in the Americas and that interference by European powers would be regarded as an act of aggression. It was later to assume greater significance. See https://history.state.gov/milestones/1801-1829/monroe

2. See Yergin, *The Prize*.

3. President Richard Nixon, 'Address to the Nation About National Energy Policy', 25 November 1973, http://www.presidency.ucsb.edu/ws/?pid=4051

4. The IEA is an autonomous body which was established in November 1974 within the framework of the OECD to implement an international energy programme. It carries out a comprehensive programme of energy cooperation among twenty-six of the thirty OECD countries. The basic aims of the IEA are: to maintain and improve systems for coping with oil supply disruptions; to promote rational energy policies in a global context through cooperative relations with non-member countries, industry and international organizations; to operate a permanent information system on the international oil market; to improve the world's energy supply and demand structure by developing alternative energy sources and increasing the efficiency of energy use; to promote international collaboration on energy technology; and to assist in the integration of environmental and energy policies. https://www.iea.org/media/aboutus/4_ieahistory.pdf

5. President Jimmy Carter, 'Solar Energy Message to the Congress', 20 June 1979, http://www.presidency.ucsb.edu/ws/?pid=32503

6. See http://www.presidency.ucsb.edu/ws/?pid=7373

7. On the problems with Carter's energy policy as seen from the political perspective of the next two decades in the US, see Wilentz, S., *The Age of Reagan: A History 1974–2008*, HarperCollins, 2008, Chapter 3.

8. US arms exports totalled $143 billion in 2011, or 6.8% of all exports. Accounting only for exports to the Middle East, that figure falls to $24 billion, or 1.1% of all exports. See: World Military Expenditure and Arms Transfers 2014, U.S. Department of State, Bureau of Verification, Compliance, and Implementation, http://www.state.gov/t/avc/rls/rpt/wmeat/2014/index.htm

9. See Knittel, 'Reducing Petroleum Consumption'.

Chapter 5: The Middle East

1. Strictly speaking it should be called the Samazov–Sykes–Picot agreement since it was an agreement between the Russians, the British and the French about the division of the Ottoman Empire between the three imperial players, with the Russians taking the lion's share. The Russian Revolution in 1917 led to a withdrawal of all its claims on the Ottoman lands, right back to the Caucasian heartlands. Thereafter, only the British and the French were left in the game, since the US made no territorial claims, and indeed President Woodrow Wilson's Fourteen Points, set out in 1918, included the right to independence for previously subjected peoples.

2. For a history of OPEC see Terzian, *OPEC*.

3. See Bamberg, J. H., *The History of the British Petroleum Company*, vol. 2, *The Anglo-Iranian Years 1928–1954*, Cambridge University Press, 2009.

4. This is a point persuasively argued in Ross, M., *The Oil Curse: How Petroleum Wealth Shapes the Development of Nations*, Princeton University Press, 2013. See also Hogan, W., Sturzenegger, F. and Tai, L., 'Contracts and Investment in Natural Resources', in Hogan, W. and Sturzenegger, F. (eds), *The Natural Resources Trap: Private Investment without Public Commitment*, MIT Press, 2010.

5. For a somewhat contrary view, see Kaufmann, R. K., Dees, S., Karadeloglou, P. and Sánchez, M., 'Does OPEC Matter? An Econometric Analysis of Oil Prices', *The Energy Journal*, 25(4), 2004, pp. 67–90. For a recent review of the econometric literature on the determination of oil prices, see Hamilton, J. D., 'Understanding Crude Oil Prices', NBER Working Paper 14492, November 2008, http://www.nber.org/papers/w14492

6. *The Economist*, 'The Next Shock?', 4 March 1999, http://www.economist.com/node/188181

7. *P5 + 1* refers to the UN Security Council's five permanent members (China, France, Russia, the UK and the US) plus Germany.

8. In the early 2000s, it was argued by Matthew Simmons that Saudi reserves were seriously overstated and that, as a result, an oil crunch would occur as Saudi's production

disappointed. Whatever the merits of the specific field assessments, there appears to have been no impediment to its subsequent production, and this was before the great technological advances in drilling later that decade. See Simmons, M. R., *Twilight in the Desert: The Coming Saudi Oil Shock and the World Economy*, John Wiley & Sons, 2005.

9. See Seznec, J-F., 'Saudi Energy Changes: The End of the Rentier State?', Atlantic Council, 24 March 2016, http://www.atlanticcouncil.org/publications/reports/saudi-energy-changes-the-end-of-the-rentier-state

10. Vision 30 is actually a rolling series of statements and documents: http://vision2030.gov.sa/en

11. The list includes tourism, travel, hospitality, retail and wholesale as the great hopes for generating employment, augmented by sending home lots of foreign workers.

12. For a description of this famous meeting on USS *Quincy*, see Lippman, T. W., 'The Day FDR met Saudi Arabia's Ibn Saud', *The Link*, 38(2), April–May 2005, http://www.ameu.org/getattachment/51ee4866-95c1-4603-b0dd-e16d2d49fcbc/The-Day-FDR-Met-Saudi-Arabia-Ibn-Saud.aspx. See also Bronson, R., *Thicker than Oil: America's Uneasy Partnership with Saudi Arabia*, Oxford University Press, 2006.

13. See Damianova, K. K., 'Iran's Re-emergence on Global Energy Markets: Opportunities, Challenges and Implications', Department of War Studies, Strategy Paper 7, EUCERS, Kings College, London, 2015, http://www.naturalgaseurope.com/iran-re-emergence-global-energy-markets-opportunities-challenges-implications-27783

14. On Syrian history, see McHugo, J., *Syria: From the Great War to Civil War*, Saqi Books, 2014.

15. See Rogan, E., *The Fall of the Ottomans: The Great War in the Middle East, 1914–1920*, Allen Lane, 2015.

Chapter 6: Russia

1. For a comprehensive history see Service, R., *A History of Modern Russia: From Nicholas II to Putin*, New Edition, Penguin, 2003.

2. See Gustafson, T., *Wheel of Fortune: The Battle for Oil and Power in Russia*, Belknap Press of Harvard University Press, 2012.

3. See Goldman, M. I., *Petrostate: Putin, Power and the New Russia*, Oxford University Press, 2008.

4. See Kotkin, S., *Armageddon Averted: The Soviet Collapse 1970–2000*, Updated Edition, Oxford University Press, 2008.

5. Brezhnev died in 1982 and there followed two further short presidencies (Yuri Andropov and Konstantin Chernenko), terminated by death in both cases.

6. Lukoil and Novatek remain private.

7. For an account of Putin's first term in office, see Shevtsova, L., *Putin's Russia*, Revised and Expanded Edition, Carnegie Endowment for International Peace, Washington, 2005; and for the period to 2008 see Stuermer, M., *Putin and the Rise of Russia*, Weidenfeld & Nicolson, 2008.

8. See Gaddy, C. and Hill, F., *Mr. Putin: Operative in the Kremlin*, Revised Edition, Brookings Institution Press, 2015.

9. Tony Blair was the first Western leader to visit Putin, whilst he was still only acting president. Bush, after meeting Putin in 2001 stated that: 'I looked the man in the eye. I found him to be very straightforward and trustworthy. We had a very good dialogue. I was able to get a sense of his soul. He's a man deeply committed to his country and the best interests of his country.' http://news.bbc.co.uk/1/hi/world/europe/1392791.stm

10. See Treisman, D., *The Return: Russia's Journey from Gorbachev to Medvedev*, Free Press (Simon & Schuster), 2011, pp. 95–6.

11. See Chapter 5 of Hoffman, D. E., *The Oligarchs: Wealth and Power in the New Russia*, Perseus Book Group, 2002.

12. The court case in 2016 questioned, among other things, the nature of the original auction of Yukos shares.

13. For more on the Yukos affair, see Sakwa, R., *Putin and the Oligarch: The Khodorkovsky–Yukos Affair*, I.B. Tauris, 2014.

14. See Browne, J., *Beyond Business: An Inspirational Memoir from a Visionary Leader*, Orion, 2011.

15. See http://www.bp.com/en_ru/russia/about-bp-in-russia/business.html

16. These are among many of the revelations about money laundering by Putin's associates revealed in the 'Panama papers'.

17. See Korostikov, M., 'Leaving to Come Back: Russian Senior Officials and State-owned Companies, Policy Papers', Russie.Nei.Visions, 87, August 2015, https://www.ifri.org/en/publications/enotes/russieneivisions/leaving-come-back-russian-senior-officials-and-state-owned

18. Dawisha, K., *Putin's Kleptocracy: Who Owns Russia?*, Simon & Schuster, 2014.

19. See Wilson, A., *Ukraine Crisis: What it Means for the West*, Yale University Press, 2014.

20. See Hosking, G., *Russia and the Russians: From Earliest Times to 2001*, Penguin, 2002.

21. See Plokhy, S., *The Gates of Europe: A History of Ukraine*, Basic Books, 2015.

22. 'Russia is effectively a non-player in robotics despite its industrial base. It neither produces nor buys robots to any significant degree, instead maintaining extractive industries (natural gas, oil, iron and nickel) and industrial manufacturing plants that look and function the way they did in the 1970s and 1980s.' Ross, A., *The Industries of the Future*, Simon & Schuster, 2016, p. 20.

23. See Monaghan, A., 'Russian State Mobilization: Moving the Country on to a War Footing', Chatham House, Russia and Eurasia Programme, May 2016, https://www.chathamhouse.org/sites/files/chathamhouse/publications/research/2016-05-20-russian-state-mobilization-monaghan-2.pdf. See especially the quotation from Putin on p. 30.

Chapter 7: China

1. See Mitter, R., *China's War with Japan 1937–1945: The Struggle for Survival*, Penguin, 2014.

2. Dyer, G., *The Contest of the Century: The New Era of Competition with China*, Allen Lane, 2014.

3. See Gittings, J., *The Changing Face of China: From Mao to Market*, Oxford University Press, 2006.

4. See Fukuyama, F., *The End of History and the Last Man*, Penguin, 1992.

5. See Chan, K.-W., 'Migration and Development in China: Trends, Geography and Current Issues', *Migration and Development*, 1(2), 2012, pp. 187–205; and National Bureau of Statistics of China, *China Statistical Yearbook 2014*, http://www.stats.gov.cn/tjsj/ndsj/2014/indexeh.htm

6. See National Bureau of Statistics of China, *China Statistical Yearbook 1996*, http://www.stats.gov.cn/english/statisticaldata/yearlydata/YB1996e/index1.htm; *China Statistical Yearbook 2014*, http://www.stats.gov.cn/tjsj/ndsj/2014/indexeh.htm; 'Statistical Communiqué of the People's Republic of China on the 2014 National Economic and Social Development', 26 February 2015, http://www.stats.gov.cn/english/PressRelease/201502/t20150228_687439.html

7. International Energy Agency, *Medium-Term Coal Market Report 2014*, OECD/IEA, 2014.

8. Indeed one of the contributing factors to the fall of Yukos was its desire to build oil pipelines to China. See Chapter 6.

9. See https://www.chathamhouse.org/publication/twt/what-exactly-one-belt-one-road and http://www.brookings.edu/research/papers/2015/07/china-regional-global-power-dollar

10. Admiral Mahan articulated in the late nineteenth century the importance of naval bases as part of the US strategy to protect trade routes. For the US's influence on China, see Hayton, B., *The South China Sea: The Struggle for Power in Asia*, Yale University Press, 2014, p. 49.

11. This has been called the 'string of pearls' strategy – including ports in Pakistan, Myanmar, Sudan, Sri Lanka, Bangladesh and the Maldives.

12. For a general survey of the multiple problems China faces, see Beardson, T., *Stumbling Giant: The Threats to China's Future*, Yale University Press, 2013.

13. There have been a number of indices constructed to reflect changes in the fundamentals. One much quoted is the Li Keqiang Index, named after a comment by him to the US Ambassador in 2007 that GDP numbers were 'man-made' (Li Keqiang is currently the Premier of the State Council of the People's Republic of China). It combines electricity consumption, rail freight volumes and bank lending, but it does not include services.

14. Nakamoto, M. and Wighton, D., 'Citigroup Chief Stays Bullish on Buy-Outs', *Financial Times*, 9 July 2007, http://www.ft.com/cms/s/0/80e2987a-2e50-11dc-821c-0000779 fd2ac.html#axzz3i3LBsxua

15. See Ansar, A., Flyvbjerg, B., Budzier, A., Lunn, D., 'Does Infrastructure Investment Lead to Economic Growth or Economic Fragility? Evidence from China', *Oxford Review of Economic Policy*, 32(3), 2016 (forthcoming).

16. Pritchett, L. and Summers, L., 'Asiaphoria Meets Regression to the Mean', NBER Working Paper No. 20573, October 2014.

17. United Nations Department of Economic and Social Affairs, Population Division, *World Population Prospects: The 2015 Revision*, United Nations, 2015.

18. There is still a strong rural link, and a great human migration back to the countryside takes place every Chinese New Year. On the wider social issues, see Lemos, G., *The End of the Chinese Dream: Why Chinese People Fear the Future*, Yale University Press, 2013.

19. Its own labour force may also be under threat from robots. See Benedikt Frey, C. and Osborne, M. A., 'The Future of Employment: How Susceptible are Jobs to Computerisation?', 17 September 2013, http://www.oxfordmartin.ox.ac.uk/downloads/ academic/The_Future_of_Employment.pdf

20. Green, F. and Stern, N., 'China's "new normal": Structural change, better growth, and peak emissions', Policy Report, Grantham Research Institute, 8 June 2015.

21. China's approach includes buying up robotic companies abroad, for example in its bid for Kuka in Germany in 2016.

Chapter 8: Europe

1. See Lieven, D., *Towards the Flame: Empire, War and the End of Tsarist Russia*, Penguin, 2015, p. 31.

2. Italy's reliance on Russia explains in part its objections to Nord Stream 2: it does not want Southern Europe cut off from the primary Russian–German engagement, especially if and when Russia terminates its exports via the Ukrainian pipelines.

3. Commission of the European Communities, 'Completing the Internal Market', White Paper from the Commission to the European Council, Com(85) 310 final, Brussels, 1985.

4. On 25 June 2009, the Council of the European Union formally adopted the new liberalization package for the European gas and electricity markets, also known as the Third Energy Package. See http://www.linklaters.com/Insights/Thirdenergypackage/Pages/ Index.aspx#sthash.n9upRh8L.dpuf

5. See Helm, D., 'European Energy Policy', in Jones, E., Menon, A. and Weatherill, S. (eds), *The Oxford Handbook of the European Union*, Oxford University Press, 2012, Chapter 39.

6. Competition and Markets Authority, 'Energy Market Investigation: Summary of Provisional Decision on Remedies', 10 March 2016.

7. At the end of the decade, in 2000, there was a Commission Green Paper on security of supply, but it was as much a concession to the national champions (which argued that

liberalization threatened security) as it was a serious attempt to change policy. Commission of the European Communities, 'Towards a European Strategy for the Security of Energy Supply', Green Paper, COM(2000) 769 final, Brussels, 29 November 2000.

8. This was in response to the Soviet deployment of SS20 missiles in Eastern Europe.

9. The Stade nuclear power station was closed in 2003, to be followed by Obrigheim in 2005.

10. See http://www.world-nuclear.org/information-library/country-profiles/countries-g-n/germany.aspx

11. See Helm, *The Carbon Crunch*, Chapter 4.

12. The Christian Democratic Union party lost badly enough in Baden-Württemberg for the Greens and the SPD to be able to form a coalition, with the Greens gaining their first regional premier.

13. Ecofys, 'Subsidies and Costs of EU Energy', Final Report, 2014, https://ec.europa.eu/energy/sites/ener/files/documents/ECOFYS%202014%20Subsidies%20and%20costs%20of%20EU%20energy_11_Nov.pdf. See also https://www.iea.org/Textbase/npsum/ElecCost2015SUM.pdf on renewables costs more generally.

14. See European Commission, 'Supply of Gas Regulation', Fact Sheet, 16 February 2016, http://europa.eu/rapid/press-release_MEMO-16-308_en.htm

15. See EurActiv, 'Russian Confirms Decision to Abandon South Stream', 10 December 2014, http://www.euractiv.com/section/energy/news/russia-confirms-decision-to-abandon-south-stream/

16. Indeed, so desperate had Germany become that, at Turkey's President Erdoğan's request, Merkel intervened to allow a German comedian to be prosecuted for mocking him.

17. In total, 7.3 GW of hard coal and 5.5 GW of lignite plants have been built in Germany since 2000, the largest of which are: RWE's 0.9 GW lignite Niederaußem plant, 2.1 GW lignite BoA 2 & 3 plants, and 1.5 GW hard coal Westfalen plants (opened in 2002, 2012, and 2014, respectively); Vattenfall's 0.9 GW and 0.6 GW lignite Boxberg plants, and 1.5 GW hard coal Moorburg plant (opened in 2000, 2012, and 2015, respectively); EnBW's 0.8 GW hard coal Rheinhafen-Dampfkraftwerk plant (opened in 2014); and Grosskraftwerk Mannheim's 0.8 GW hard coal GKM plant (opened in 2015).

18. The numbers by summer 2016 are roughly; debts, €37 billion; existing reactor upgrade costs, €55 billion; Areva, €2–€5 billion; and its share of Hinkley, €16 billion.

19. Indeed Europe experienced higher temperatures in the medieval warm period, without obvious detriments. The Vikings got to Iceland, Greenland and even North America, and managed to sustain settlements precisely because it was warmer. Even before this the Romans got so far north, and even cultivated vineyards in York, because of a warmer climate. See Helm, *The Carbon Crunch*, pp. 29–30.

Chapter 9: The gradual end of Big Oil

1. For a list see Helman, C., 'The World's Biggest Oil and Gas Companies – 2015', Forbes, 19 March 2015, http://www.forbes.com/sites/christopherhelman/2015/03/19/the-worlds-biggest-oil-and-gas-companies/#3af094a6b68d. See also Thurber, M., 'NOCs and the Global Oil Market: Should We Worry?', Program on Energy and Sustainable Development, Stanford University, 6 February 2012, https://energy.stanford.edu/sites/default/files/thurber_energy_seminar_nocs_06feb2012_final_0.pdf

2. There are of course exceptions. One of the world's most recognizable automobile brands, Rolls-Royce, is arguably as famous today for its aircraft engines as its cars, but the key point is that Ford and General Motors, while they have had aviation divisions, are not.

3. See Shell, 'Shell to Halt Carmon Creek In Situ Project', Media Release, 27 October 2015, http://www.shell.com/

4. See Doran, P. B., *Breaking Rockefeller: The Incredible Story of the Ambitious Rivals Who Toppled an Oil Empire*, Viking, 2016.

5. The Anglo-Iranian Oil Company's concession was nationalized in 1951 and passed to the National Iranian Oil Company (NIOC). After the 1953 coup to remove Mosaddegh, the concession was renegotiated and the 'contractors', BP and the US companies, were given 40% and 60% shares, respectively, while NIOC remained the executive owner of the oil resources. In 1973, the Shah negotiated a new 20-year agreement with the foreign companies, and the government passed a law that granted NIOC formal control over Iran's oil assets. In 1979, after the revolution, all oil agreements were cancelled and NIOC took control of oil operations.

6. See Bamberg, J. H., *British Petroleum and Global Oil 1950–75: The Challenge of Nationalism*, Cambridge University Press, 2000.

7. See Winter, D. A. and King, B., 'The West Sole Field, Block 48/6, UK North Sea', Geological Society, London, *Memoirs* 14(1), 1991, pp. 517–23. See also Kemp, A., *The Official History of North Sea Oil and Gas*, 2 vols, Routledge, 2012.

8. See Hinton, D. D., 'The Seventeen-Year Overnight Wonder: George Mitchell and Unlocking the Barnett Shale', *Journal of American History*, 99(1), 2012, pp. 229–35, doi: 10.1093/jahist/jas064

9. The problem with this merger is not the strategy but the price. It needed an oil price of $60–$70 to justify it.

10. One possible rule is that assets in aggregate should be maintained and enhanced to ensure that the interests of future generations are taken into account. See Helm, *Natural Capital*, Chapter 3.

11. See Browne, J., *Beyond Business: An Inspirational Memoir from a Visionary Leader*, Orion, 2011.

12. See Helm, D., 'What should oil companies do about climate change?', 26 February 2015, www.dieterhelm.co.uk

Chapter 10: Energy utilities

1. See Helm, D., *Energy, the State and the Market: British Energy Policy Since 1979*, Revised Edition, Oxford University Press, 2004.

2. Hicks states that, 'The best of all monopoly profits is a quiet life.' Hicks, J. R., 'Annual Survey of Economic Theory: The Theory of Monopoly', *Econometrica*, 3(1), January 1935, p. 8.

3. Helm, *Energy, the State and the Market*.

4. See Newbery, D., *Privatization, Restructuring, and Regulation of Network Utilities*, The Walras-Pareto Lectures, MIT Press, 1999.

5. For an explanation of Ramsey pricing and its application, see Wolak, F. A., 'Public utility pricing and finance', in Durlauf, S. N. and Blume, L. E., *The New Palgrave Dictionary of Economics*, Second Edition, 2008, http://web.stanford.edu/group/fwolak/cgi-bin/sites/default/files/wolak_palgrave.pdf

6. See Helm, *Energy, the State and the Market*, pp. 133 and 308–9.

7. This is called the 'missing money' problem. See Newbery, D., 'Missing Money and Missing Markets: Reliability, Capacity Auctions and Interconnectors', EPRG Working Paper 1508, Cambridge Working Paper in Economics 1513, 2015.

8. RWE and E.ON were partners in the Horizon venture to build new nuclear in Britain, starting at Wylfa in North Wales. They pulled out in 2012, and Hitachi took over the company. This followed an earlier enthusiasm in the mid-2000s to build new large coal-fired power stations, and again both separately backed out.

9. See *Sunday Telegraph*, 'Ofgem Chief Dermot Nolan: "The industry now has got to deliver"', 28 May 2016, http://www.telegraph.co.uk/business/2016/05/28/ofgem-chief-dermot-nolan-the-industry-now-has-got-to-deliver/

10. See http://www.economist.com/news/business/21582018-sustainable-energy-meets-unsustainable-costs-cost-del-sol

11. Google, Toyota and Honda lead the corporate world in investing in R&D on robotics. See Ross, A., *The Industries of the Future*, Simon & Schuster, 2016, Chapter 1.
12. See Conn, I., 'Strategic Review', Centrica, July 2015, https://www.centrica.com/sites/default/files/transcript-iain-conn-strategic-review-group_overview.pdf
13. Enron started out as a conventional utility, as the result of a merger of two gas pipeline companies, Houston Natural Gas and InterNorth.

Chapter 11: The new energy markets and the economics of the Internet

1. It is revealing that Hubbert (of peak oil fame) was an early member of the 'technocracy' movement in the US in the 1930s. See part two of Inman, M., *The Oracle of Oil: The Maverick Geologist's Quest for a Sustainable Future*, W. W. Norton, 2016.
2. See Frunza, M.-C., *Fraud and Carbon Markets: The Carbon Connection*, Environmental Market Insights, Routledge, 2015.
3. See Fattouh, B., Kilian, L. and Mahadeva, L., 'The Role of Speculation in Oil Markets: What Have We Learned so Far?', CEPR Discussion Paper No. DP8916, Centre for Economic Policy Research (CEPR), March 2012.
4. See Schiller, R. J., *Irrational Exuberance*, Revised and Expanded Third Edition, Princeton University Press, 2015; and Reinhart, C. M. and Rogoff, K. S., *This Time is Different: Eight Centuries of Financial Folly*, Princeton University Press, 2009.
5. 'It is not a case of choosing those [faces] which, to the best of one's judgment, are really the prettiest, nor even those which average opinion genuinely thinks the prettiest. We have reached the third degree where we devote our intelligences to anticipating what average opinion expects the average opinion to be. And there are some, I believe, who practise the fourth, fifth and higher degrees.' See Keynes, J. M., *The General Theory of Employment Interest and Money*, Macmillan, 1936.
6. For this reason, it makes little sense to offer fixed-price contracts linked to the wholesale price. Yet that is exactly what the British FiTs do, and others as well.

Conclusion

1. Gordon, R. J., *The Rise and Fall of American Growth: The US Standard of Living Since the Civil War*, Princeton University Press, 2016.

Bibliography

Abraham, D. S., *The Elements of Power: Gadgets, Guns and the Struggle for a Sustainable Future in the Rare Metal Age*, Yale University Press, 2016

Alekperov, V., *Oil of Russia: Past, Present and Future*, Lukoil, Minneapolis: East View Press, December 2010, pp. 150–151, http://www.litasco.com/_library/pdf/media/oil_of_russia_by_Vagit_ALEKPEROV_full%20edition.pdf

Allen, M., 'Drivers of Peak Warming in a Consumption-maximising World', *Nature Climate Change*, May 2016, http://www.nature.com/nclimate/journal/vaop/ncurrent/full/nclimate2977.html

Allen, M., Frame, D., Huntingford, C., Jones, C., Lowe, J., Meinshausen, M. and Meinshausen, N., 'Warming Caused by Cumulative Carbon Emissions Towards the Trillionth Tonne', *Nature*, 458(7242), 2009, pp. 1163–6

Ansar, A., Flyvbjerg, B., Budzier, A., Lunn, D., 'Does Infrastructure Investment Lead to Economic Growth or Economic Fragility? Evidence from China', *Oxford Review of Economic Policy*, 32(3), 2016 (forthcoming)

Aurora Energy Research, 'Predictable Surprises: Lessons from 30 Years of Energy Sector Forecasts', 2013

Bamberg, J. H., *British Petroleum and Global Oil 1950–75: The Challenge of Nationalism*, Cambridge University Press, 2000

Bamberg, J. H., *The History of the British Petroleum Company*, vol. 2, *The Anglo-Iranian Years 1928–1954*, Cambridge University Press, 2009

Beardson, T., *Stumbling Giant: The Threats to China's Future*, Yale University Press, 2013

Blas, J. and Flood, C., 'Analyst Warns of Oil at $200 a Barrel', *Financial Times*, 6 May 2008, http://www.ft.com/cms/s/0/70b4ef0a-1b91-11dd-9e58-0000779fd2ac.html?siteedition=uk#axzz3igHrWU9r

Brandon, N. P. et al., 'UK Research Needs in Grid Scale Energy Storage Technologies', http://energysuperstore.org/wp-content/uploads/2016/04/IMPJ4129_White_Paper_UK-Research-Needs-in-Grid-Scale-Energy-Storage-Technologies_WEB.pdf

Bronson, R., *Thicker than Oil: America's Uneasy Partnership with Saudi Arabia*, Oxford University Press, 2006

Browne, J., *Beyond Business: An Inspirational Memoir from a Visionary Leader*, Orion, 2011

Brynjolfsson, E. and McAfee, A., *The Second Machine Age: Work, Progress, and Prosperity in a Time of Brilliant Technologies*, W. W. Norton, 2016

Campbell, C. and Laherrere, J., 'The End of Cheap Oil', *Scientific American*, March 1998

Chan, K-W., 'Migration and Development in China: Trends, Geography and Current Issues', *Migration and Development*, 1(2), 2012, pp. 187–205

Coady, D., Parry, I., Sears, L. and Shang, B., 'How Large are Global Energy Subsidies?', IMF Working Paper WP/15/105, 2015

Commission of the European Communities, 'Completing the Internal Market', White Paper from the Commission to the European Council, Com(85) 310 final, Brussels, 1985

Commission of the European Communities, 'Towards a European Strategy for the Security of Energy Supply', Green Paper, COM(2000) 769 final, Brussels, 29 November 2000

Competition and Markets Authority, 'Energy Market Investigation: Summary of Provisional Decision on Remedies', 10 March 2016

Conn, I., 'Strategic Review', Centrica, July 2015, https://www.centrica.com/sites/default/files/transcript-iain-conn-strategic-review-group_overview.pdf

Damianova, K. K., 'Iran's Re-emergence on Global Energy Markets: Opportunities, Challenges and Implications', Department of War Studies, Strategy Paper 7, EUCERS, Kings College, London, 2015, http://www.naturalgaseurope.com/iran-re-emergence-global-energy-markets-opportunities-challenges-implications-27783

Davis, S. C., Diegel, S. W. and Boundy, R. G., *Transportation Energy Data Book*, Edition 33, Oak Ridge National Laboratory, July 2014, cta.ornl.gov/data

Dawisha, K., *Putin's Kleptocracy: Who Owns Russia?*, Simon & Schuster, 2014

Delbecke, J. and Vis, P. (eds), *EU Climate Policy Explained*, Routledge, 2015

Department of Energy & Climate Change, 'Updated Energy and Emissions Projections 2014', 25 September 2014, https://www.gov.uk/government/uploads/system/uploads/attachment_data/file/368021/Updated_energy_and_emissions_projections2014.pdf

Diegel, S. W., *Transportation Energy Data Book*, Edition 33, Oak Ridge National Laboratory, July 2014, cta.ornl.gov/data

Doran, P. B., *Breaking Rockefeller: The Incredible Story of the Ambitious Rivals Who Toppled an Oil Empire*, Viking, 2016

Dyer, G., *The Contest of the Century: The New Era of Competition with China*, Allen Lane, 2014

Ecofys, 'Subsidies and Costs of EU Energy', Final Report, 2014, https://ec.europa.eu/energy/sites/ener/files/documents/ECOFYS%202014%20Subsidies%20and%20costs%20of%20EU%20energy_11_Nov.pdf

Energy Information Administration, 'Monthly Energy Review', April 2016, http://www.eia.gov/totalenergy/data/monthly/archive/00351604.pdf

Energy Information Administration, 'Natural gas production by source in the Reference case, 1990–2040', *Natural Gas Annual 2012*, DOE/EIA-0131 (2012), Washington, DC, December 2013

Fagerberg, E. J., Mowery, D. and Nelson, R. (eds), *The Oxford Handbook of Innovation*, Oxford University Press, 2005

Fattouh, B., Kilian, L. and Mahadeva, L., 'The Role of Speculation in Oil Markets: What Have We Learned so Far?', CEPR Discussion Paper No. DP8916, Centre for Economic Policy Research (CEPR), March 2012

Ford, M., *The Rise of the Robots: Technology and the Threat of Mass Unemployment*, Oneworld Publications, 2015

Frunza, M-C., *Fraud and Carbon Markets: The Carbon Connection*, Environmental Market Insights, Routledge, 2015

Fukuyama, F., *The End of History and the Last Man*, Penguin, 1992

Gaddy, C. and Hill, F., *Mr. Putin: Operative in the Kremlin*, Revised Edition, Brookings Institution Press, 2015

Gately, D., 'Lessons from the 1986 Oil Price Collapse', *Brookings Papers on Economic Activity* 2, 1986, http://www.brookings.edu/~/media/Projects/BPEA/1986-2/1986b_bpea_gately_adelman_griffin.PDF

Gittings, J., *The Changing Face of China: From Mao to Market*, Oxford University Press, 2006

Goldman, M. I., *Petrostate: Putin, Power and the New Russia*, Oxford University Press, 2008

Gordon, R. J., *The Rise and Fall of American Growth: The US Standard of Living Since the Civil War*, Princeton University Press, 2016

Graetz, M. J., *The End of Energy: The Unmaking of America's Environment, Security and Independence*, MIT Press, 2011

Green, F. and Stern, N., 'China's "new normal": Structural change, better growth, and peak emissions', Policy Report, Grantham Research Institute, 8 June 2015

Gustafson, T., *Wheel of Fortune: The Battle for Oil and Power in Russia*, Belknap Press of Harvard University Press, 2012

Hamilton, J. D., 'The Changing Face of World Oil Markets', NBER Working Paper 20355, July 2014, http://www.nber.org/papers/w20355

Hamilton, J. D., 'Understanding Crude Oil Prices', NBER Working Paper 14492, November 2008, http://www.nber.org/papers/w14492

Haszeldine, R. S., 'Can CCS and NET enable the continued use of fossil carbon fuels after COP21?', p. 304, in Helm, D. (ed.), *The Future of Fossil Fuels: Oxford Review of Economic Policy*, 32(2), Summer 2016, www.oxrep.oxfordjournals.org

Hayton, B., *The South China Sea: The Struggle for Power in Asia*, Yale University Press, 2014

Helm, D., *The Carbon Crunch: How We're Getting Climate Change Wrong – And How to Fix It*, revised edition, Yale University Press, 2013

Helm, D., *Energy, the State and the Market: British Energy Policy Since 1979*, Revised Edition, Oxford University Press, 2004

Helm, D., 'EU Climate-Change Policy – A Critique', in Helm, D. R. and Hepburn, C. (eds), *The Economics and Politics of Climate Change*, Oxford University Press, 2009

Helm, D., 'European Energy Policy', in Jones, E., Menon, A. and Weatherill, S. (eds), *The Oxford Handbook of the European Union*, Oxford University Press, 2012, Chapter 39

Helm, D. (ed.), *The Future of Fossil Fuels: Oxford Review of Economic Policy*, 32(2), Summer 2016, www.oxrep.oxfordjournals.org

Helm, D., *Natural Capital: Valuing the Planet*, Yale University Press, 2015

Helm, D., 'What should oil companies do about climate change?', 26 February 2015, www.dieterhelm.co.uk

Helm, D., Phillips, J. and Smale, R., 'Too Good to be True? The UK's Climate Change Record', 2008, www.dieterhelm.co.uk

Helman, C., 'The World's Biggest Oil and Gas Companies – 2015', *Forbes*, 19 March 2015, http://www.forbes.com/sites/christopherhelman/2015/03/19/the-worlds-biggest-oil-and-gas-companies/#3af094a6b68d

Hepburn, C., 'Regulating by Price, Quantities or Both: An Update and an Overview', *Oxford Review of Economic Policy*, 22(2), 2006, pp. 226–247

Hicks, J. R., 'Annual Survey of Economic Theory: The Theory of Monopoly', *Econometrica*, 3(1), January 1935

Hinton, D. D., 'The Seventeen-Year Overnight Wonder: George Mitchell and Unlocking the Barnett Shale', *Journal of American History*, 99(1), 2012, pp. 229–235, doi: 10.1093/jahist/jas064

Hoffman, D. E., *The Oligarchs: Wealth and Power in the New Russia*, Perseus Book Group, 2002

Hogan, W., Sturzenegger, F. and Tai, L., 'Contracts and Investment in Natural Resources', in Hogan, W. and Sturzenegger, F. (eds), *The Natural Resources Trap: Private Investment without Public Commitment*, MIT Press, 2010

Hosking, G., *Russia and the Russians: From Earliest Times to 2001*, Penguin, 2002

Howarth, S., *A Century in Oil: The 'Shell' Transport and Trading Company 1987–1997*, Weidenfeld & Nicolson, 1997

Hubbert, M., 'The Energy Resources of the Earth', *Scientific American*, 225, September 1971, pp. 60–70

Inman, M., *The Oracle of Oil: The Maverick Geologist's Quest for a Sustainable Future*, W. W. Norton, 2016

International Energy Agency, 'CO$_2$ Emissions from Fuel Combustion: Highlights', 2011, http://www.iea.org/media/statistics/co2highlights.pdf

International Energy Agency, 'IEA Headline Global Energy Data', 2015, http://www.iea.org/statistics/

International Energy Agency, 'Medium-term Coal Market Report 2014', OECD/IEA, 2014

International Energy Agency, 'Medium Term Coal Outlook: Market Analysis and Forecasts to 2020', IEA, 2015

International Energy Agency, 'World CO$_2$ Emissions from Fuel Combustion', OECD/IEA, 2015, http://wds.iea.org/wds/pdf/Worldco2_Documentation.pdf

International Energy Agency, *World Energy Outlook*, 1982, 1993, 2000, 2006, 2010, 2012, 2013, 2014, 2015, OECD

Jevons, W. S., *The Coal Question: An Inquiry Concerning the Progress of the Nation, and the Probable Exhaustion of Our Coal Mines*, Dodo Press, 2008

Johnston, L. and Williamson, S. H., 'What Was the U.S. GDP Then?', MeasuringWorth, http://www.measuringworth.org/usgdp/

Kam, W. C., 'Migration and Development in China: Trends, Geography and Current Issues', *Migration and Development*, 1(2), 2012, pp. 187–205

Kemp, A., *The Official History of North Sea Oil and Gas*, 2 vols, Routledge, 2012

Keohane, R. O. and Victor, D. G., 'Cooperation and Discord in Global Climate Policy', *Nature Climate Change: Perspective*, 6, 9 May 2016, pp. 570–5

Knittel, C. R., 'Reducing Petroleum Consumption from Transportation', *Journal of Economic Perspectives*, 26(1), Winter 2012, pp. 93–118

Korostikov, M., 'Leaving to Come Back: Russian Senior Officials and State-owned Companies, Policy Papers', Russie.Nei.Visions, 87, August 2015, https://www.ifri.org/en/publications/enotes/russieneivisions/leaving-come-back-russian-senior-officials-and-state-owned

Kotkin, S., *Armageddon Averted: The Soviet Collapse 1970–2000*, Updated Edition, Oxford University Press, 2008

Lemos, G., *The End of the Chinese Dream: Why Chinese People Fear the Future*, Yale University Press, 2013

Lester, R. K. and Hart, D. M., *Unlocking Energy Innovation: How America Can Build a Low-cost Low-carbon Energy System*, MIT Press, 2012

Leveque, F., *The Economics and Uncertainties of Nuclear Power*, Cambridge University Press, 2015

Lieven, D., *Towards the Flame: Empire, War and the End of Tsarist Russia*, Penguin, 2015

Lippman, T. W., 'The Day FDR met Saudi Arabia's Ibn Saud', *The Link*, 38(2), April–May 2005, http://www.ameu.org/getattachment/51ee4866-95c1-4603-b0dd-e16d2d49fcbc/The-Day-FDR-Met-Saudi-Arabia-Ibn-Saud.aspx

MacKay, D. J. C., *Sustainable Energy: Without the Hot Air*, UIT Cambridge, 2009

Mayer-Schönberger, V. and Cukier, K., *Big Data: A Revolution that Will Transform How We Live, Work and Think*, John Murray, 2013

McHugo, J., *Syria: From the Great War to Civil War*, Saqi Books, 2014

McKane, A., Price, L. and de la Rue du Can, S., 'Policies for Promoting Industrial Energy Efficiency in Developing Countries and Transition Economies', Background Paper for the United Nations Industrial Development Organization (UNIDO) Side Event on Sustainable Industrial Development on 8 May 2007 at the Commission for Sustainable Development (CSD-15), p. 9

Meadows, D. H., Meadows, D. L., Randers, J. and Behrens, III, W. W., *The Limits to Growth: A Report for the Club of Rome's Project on the Predicament of Mankind*, New American Library, 1972

Millar, R., Allen, M., Rogelj, J. and Friedlingstein, P., 'The Cumulative Carbon Budget and its Implications', *Oxford Review of Economic Policy*, 32(2), Summer 2016, pp. 323–42

Mitter, R., *China's War with Japan 1937–1945: The Struggle for Survival*, Penguin, 2014

Monaghan, A., 'Russian State Mobilization: Moving the Country on to a War Footing', Chatham House, Russia and Eurasia Programme, May 2016, https://www.chathamhouse. org/sites/files/chathamhouse/publications/research/2016-05-20-russian-state-mobilization-monaghan-2.pdf

Moore, G. E., 'Cramming More Components onto Integrated Circuits', Electronics Magazine, 1965

Moore, G.E., 'Progress in Digital Integrated Electronics', IEEE, IEDM Tech Digest, 1975

Nakamoto, M. and Wighton, D., 'Citigroup Chief Stays Bullish on Buy-Outs', Financial Times, 9 July 2007, http://www.ft.com/cms/s/0/80e2987a-2e50-11dc-821c-0000779fd2ac. html#axzz3i3LBsxua

National Bureau of Statistics of China, China Statistical Yearbook 1996, http://www.stats.gov. cn/english/statisticaldata/yearlydata/YB1996e/index1.htm

National Bureau of Statistics of China, China Statistical Yearbook 2014, http://www.stats.gov. cn/tjsj/ndsj/2014/indexeh.htm

National Bureau of Statistics of China, China Statistical Yearbook 2015, http://www.stats.gov. cn/tjsj/ndsj/2015/indexeh.htm

National Bureau of Statistics of China, 'Statistical Communiqué of the People's Republic of China on the 2014 National Economic and Social Development', 26 February 2015, http:// www.stats.gov.cn/english/PressRelease/201502/t20150228_687439.html

Newbery, D., 'Missing Money and Missing Markets: Reliability, Capacity Auctions and Interconnectors', EPRG Working Paper 1508, Cambridge Working Paper in Economics 1513, 2015

Newbery, D., Privatization, Restructuring, and Regulation of Network Utilities, The Walras-Pareto Lectures, MIT Press, 1999

Nickell, S., Redding, S. and Swaffield, J., 'The Uneven Pace of Deindustrialisation in the OECD', The World Economy, 2008

Nielsen, C. P. and Ho, M. S., 'Air Pollution and Health Damages in China: An Introduction and Review', in Nielsen, C. P. and Ho, M. S. (eds), Clearing the Air: The Health and Economic Damages of Air Pollution in China, MIT Press, 2007

Organization of Petroleum Exporting Countries, 'Annual Report 2014', 2015, http://www. opec.org/opec_web/static_files_project/media/downloads/publications/Annual_ Report_2014.pdf

Organization of Petroleum Exporting Countries, 'Monthly Oil Market Report', 13 May 2016, http://www.opec.org/opec_web/static_files_project/media/downloads/publications/ MOMR%20May%202016.pdf

Ostrovsky, A., The Invention of Russia: The Journey from Gorbachev's Freedom to Putin's War, Atlantic Books, 2015

Pazos-Outón, L. M., Szumilo, M., Lamboll, R., Richter, J. M., Crespo-Quesada, M., Abdi-Jalebi, M., Beeson, H. J., Vrućinić, M., Alsari, M., Snaith, H. J., Ehrler, B., Friend, R. H., and Deschler, F., 'Photon recycling in lead iodide perovskite solar cells', Science, 351, 2016, pp. 1430–3

Peters, G. P., Minx, J. C, Weber, C. L. and Edenhofer, O, 'Growth in Emission Transfers Via International Trade from 1990 to 2008', Proceedings of the National Academy of Sciences of the United States of America, 108(21), 24 May 2011, pp. 8903–8, http://www.ncbi.nlm.nih. gov/pubmed/21518879

Pfeiffer, A., Millar, R., Hepburn, C. and Beinhocker, E., 'The "2°C Capital Stock" for Electricity Generation: Cumulative Committed Carbon Emissions and Climate Change', Applied Energy, 2016

Ploeg, F. van der, 'Natural Resources: Curse or Blessing', Journal of Economic Literature, 49(2), 2011, pp. 366–420

Ploeg, F. van der, 'Race to Burn the Last Ton of Carbon and the Risk of Stranded Assets', Research Paper 178, Oxford Centre for the Analysis of Resource Rich Economics, Department of Economics, University of Oxford, May 2016

Plokhy, S., The Gates of Europe: A History of Ukraine, Basic Books, 2015

Pritchett, L. and Summers, L., 'Asiaphoria Meets Regression to the Mean', NBER Working Paper No. 20573, October 2014

Radioactive Substances Advisory Committee, Panel on Disposal of Radioactive Wastes Cmnd, 'The Control of Radioactive Wastes', HMSO, 1959

Reinhart, C. M. and Rogoff, K. S., *This Time is Different: Eight Centuries of Financial Folly*, Princeton University Press, 2009

REN21 Steering Committee, 'Renewables 2016: Global Status Report', http://www.ren21. net/wp-content/uploads/2016/06/GSR_2016_Full_Report_REN21.pdf

Rogan, E., *The Fall of the Ottomans: The Great War in the Middle East, 1914–1920*, Allen Lane, 2015

Ross, A., *The Industries of the Future*, Simon & Schuster, 2016

Ross, M., *The Oil Curse: How Petroleum Wealth Shapes the Development of Nations*, Princeton University Press, 2013

Saddy, F. (ed.), *Arab–Latin American Relations: Energy, Trade, and Investment*, Transaction, 1983

Sakwa, R., *Putin and the Oligarch: The Khodorkovsky–Yukos Affair*, I. B. Tauris, 2014

Satsangi, K. P. and Sandeepkumar, K., 'Advanced Technologies in Power Transmission System', IET-UK Conference on Information and Communication Technology in Electrical Sciences, 2007

Schiller, R. J., *Irrational Exuberance*, Revised and Expanded Third Edition, Princeton University Press, 2015

Service, R., *A History of Modern Russia: From Nicholas II to Putin*, New Edition, Penguin, 2003

Seznec, J-F., 'Saudi Energy Changes: The End of the Rentier State?', Atlantic Council, 24 March 2016, http://www.atlanticcouncil.org/publications/reports/saudi-energy-changes-the-end-of-the-rentier-state

Shevtsova, L., *Putin's Russia*, Revised and Expanded Edition, Carnegie Endowment for International Peace, Washington, 2005

Shultz, G. P. and Armstrong, R. C., *Game Changers: Energy on the Move*, Hoover Institute Press, Stanford University, 2014

Simmons, M. R., *Twilight in the Desert: The Coming Saudi Oil Shock and the World Economy*, John Wiley & Sons, 2005

Stuermer, M., *Putin and the Rise of Russia*, Weidenfeld & Nicolson, 2008

Sunday Telegraph, 'Ofgem Chief Dermot Nolan: "The industry now has got to deliver"', 28 May 2016, http://www.telegraph.co.uk/business/2016/05/28/ofgem-chief-dermot-nolan-the-industry-now-has-got-to-deliver/

Swanson, R., 'A Vision for Crystalline Silicon Photovoltaics', *Progress in Solar Photovoltaics*, 14(5), August 2006, pp. 443–53

Terzian, P., *OPEC: The Inside Story*, translated by Michael Pallis, Zed Books, 1985

The Economist, 'The Next Shock?', 4 March 1999, http://www.economist.com/node/188181

The Scottish Government, 'Scotland's Future: Your Guide to an Independent Scotland', 26 November 2013

Thurber, M., 'NOCs and the Global Oil Market: Should We Worry?', Program on Energy and Sustainable Development, Stanford University, 6 February 2012, https://energy.stanford.edu/sites/default/files/thurber_energy_seminar_nocs_06feb2012_final_0.pdf

Tol, R. J., 'The road from Paris: International climate policy after the Paris agreement of 2015', VOX CEPR Policy Portal, 17 December 2015, http://voxeu.org/article/road-cop21

Treisman, D., *The Return: Russia's Journey from Gorbachev to Medvedev*, Free Press (Simon & Schuster), 2011

Tussing, A. R. and Erickson, G. K., 'Reflections on the End of the OPEC Era', *Alaska Review of Social and Economic Conditions*, 19(4), December 1982, p. 3, http://www.iser.uaa.alaska.edu/Publications/formal/arsecs/ARSEC_XIX_4_End_of_OPEC_Era.pdf

United Nations Department of Economic and Social Affairs, Population Division, *World Population Prospects: The 2015 Revision*, United Nations, 2015

Wilentz, S., *The Age of Reagan: A History 1974–2008*, HarperCollins, 2008

Wilson, A., *Ukraine Crisis: What it Means for the West*, Yale University Press, 2014

Winter, D. A. and King, B., 'The West Sole Field, Block 48/6, UK North Sea', Geological Society, London, *Memoirs* 14(1), 1991, pp. 517–23

Wolak, F. A., 'Public utility pricing and finance', in Durlauf, S. N. and Blume, L. E., *The New Palgrave Dictionary of Economics*, Second Edition, 2008, http://web.stanford.edu/group/fwolak/cgi-bin/sites/default/files/wolak_palgrave.pdf

Yergin, D., *The Prize: The Epic Quest for Oil, Money, and Power*, Simon & Schuster, 1990

Index

Giscard d'Estaing, Valéry 23,
   249 n10
glasnost 131, 146
Glencore 225
Golan Heights 113
Gold Standard 94
Goldman Sachs 16
Google
   cars and 79, 217
   D-wave 81
   lobbying against oil companies 85
   search engines 77
   takes hold of the future 10
   transport interests 105
Gorbachev, Mikhail
   Berlin Wall and 139
   complexities of 130–1
   demonstrators and 146
   Khodorkovsky 133
   low oil prices affect 6
   Putin and 140
   Ukraine fails to support 137
Gordon, Robert J. 241–3
Gosplan 129, 145
graphene 9, 79–80, 83
'Great Leap Forward' 145
'Great Migration' 147
Great Recession 103, 175
Greece 173
Green Party (Germany) 167–8, 172, 174,
   215
greenhouse gases 40, 54, 250 n2
Greenspan, Alan 231
Grenada 101
Gulf of Mexico 29, 91, 192
Gulf States 22, 90, 98
Gulf Wars 26, 94, 98, 99, 115

Hainan Island 154
Halliburton 189
Hamilton, James 16, 248 n2 (Chapter 1)
Hezbollah 100, 117
Hicks, John 207
Hinkley Point xii, 51, 174, 215, 229
Hitler, Adolf 129, 130, 142, 145
Holland 163, 231
Homs 124
Hong Kong 146
Hormuz, Strait of 4, 153, 154
Houthis 117
Hubbert, M. King 24, 30, 33, 96, 250 n16,
   259 n1 (Chapter 11)
Humber Estuary 49
Hungary 169

Hussein, Saddam 26, 98, 100, 107
hydro power 50, 213

Iberdrola 211, 219
IBM 10, 105
Ibn Saud 102, 107
Iceland 7
India
   China and 150
   competitive advantages become less
      relevant 8
   electricity demand of 48
   fossil fuel boom in 46
   growth forecast 27
   growth plans 199
   Kyoto and 41
   nuclear power and electricity 51
Indian Ocean 153, 154, 159, 244
Indonesia 47, 112, 157, 159
Industrial Revolution 163
industrialization 28, 128
inflation 22, 26, 94
interconnectivity 73–4
International Energy Agency (IEA) xii, 3,
   17–18, 23, 253 n4 (Chapter 4)
International Monetary Fund (IMF) 22
International Power 220
Internet 7, 8, 10, 232
Internet of Things 6, 36
IOCs (international oil companies)
   countries of origin 191
   limits to diversification 202
   NOCs and 2, 183, 186, 198, 200
   share prices 184
iPhones 102
Iran 123–4
   Anglo-Persian Oil 191
   Carter embargos oil 96
   challenged by Iraq 94, 98–9, 100,
      108
   exports oil to China 151
   Gulf States and 22
   hatred of British 189
   Iraq comes under influence of 115
   Mosaddegh overthrown 185
   national oil companies 2, 21, 110–11
   1950s instability 93
   1979 Revolution 17, 94, 107, 111, 114
   oil supplies 4, 34, 117–19
   OPEC 112
   opposes Saudis 95, 107, 112, 116, 120,
      123, 243
   pipelines 171
   promising prospects 11